# Past and Current Models in Electrochemistry

Alfred Anderson

# Past and Current Models in Electrochemistry

Springer

Alfred Anderson
Department of Chemistry
Case Western Reserve University
Cleveland, OH, USA

ISBN 978-3-031-97790-9 ISBN 978-3-031-97791-6 (eBook)
https://doi.org/10.1007/978-3-031-97791-6



This Springer imprint is published by the registered company Springer Nature Switzerland AG
The registered company address is: Gewerbestrasse 11, 6330 Cham, Switzerland

*To*
*The curious*

# Preface

This monograph covers many of the key experimental and theoretical developments in electrochemistry from the 1800s to the 2020s. The goal is to give the reader a view of the historical beginnings of electrochemical science and the significant experimental and theoretical developments leading to our present understanding. It traces the computational and model approaches used today back to their early intellectual foundations. Historical and present day quantum chemical theories used to understand electrocatalysis are emphasized. Along the way are suggestions about alternative models and further computational exploration.

In today's world, electrochemistry plays an increasingly important role in the design of efficient, environmentally compatible methods for energy generation, storage, and conversion and for efficient synthesis of useful materials. Through new experimental and theoretical tools, electrochemical science has evolved from its historical basis dependent on electrochemical measurements. New spectroscopies are applied to make in situ measurements of the chemistry on electrode surfaces and its dependence on applied potential. Predictions of electrochemical reaction potentials and energies are now made using quantum chemical computations. Analyses and understanding are gained from the computations using concepts of chemical bonding theory. The author became interested in theories for electrocatalysis in 1980 at the urging of Ernest Yeager, an experimental electrochemist who directed the Case Center of Electrochemical Sciences. Approaching the field from his background in applied quantum theory, the author analyzes many of the recent advancements by theoretical and experimental researchers. The result is this introduction to electrochemistry and electrocatalysis theory that focusses phenomenological understanding as well as the mathematical expressions. It should be suitable as a reference for electrochemists who want to brush up on electrochemistry history and learn about the evolving contributions of quantum chemistry. It provides overview for theorists unfamiliar with electrochemistry but attracted to it as a field to explore.

Electrochemistry began with electrodes, solutions containing ions (electrolytes), and the determination of electrical current available through wires connecting the electrodes. Alessandro Giuseppe Antonio Anastasio Volta pioneered making electrical batteries producing electricity with zinc anodes, copper cathodes and sulfuric acid electrolyte. In his "voltaic pile" of 1800, zinc was oxidized at the anode releasing electrons which went through a wire to the copper cathode where protons in the acid were reduced to hydrogen gas. Other experimentalists continued in the nineteenth century, the most notable in the author's mind being Michael Faraday. He discovered the concept of electrical and chemical equivalents and gave us much of the language of electrochemical science used today. Well known to electrochemists is Walther Nernst who introduced the famous "Nernst equation" in his habilitation thesis of 1889 which related electrochemical cell potentials to electrolyte compositions at equilibrium. For his advancements in thermochemistry, including development of the third law of thermodynamics, he was rewarded the 1920 Nobel Prize in chemistry.

J. Willard Gibbs showed in 1876–1878 that electrical work produced by a cell is proportional to the potential and the amount of charge passed. He developed the science of statistical mechanics and showed the role of densities of states in determining thermodynamic properties. He defined what is now called the Gibbs or free energy, which is the maximum work that can be extracted from a system undergoing change; in electrochemistry the Gibbs energy is the maximum available electrical work. Through the insights of Gibbs, electrochemistry became a branch of physical chemistry.

It wasn't until 1897 that J. J. Thompson discovered that electricity, viewed as a fluid, consisted of particles, namely electrons. In the first quarter of the 20th century kinetics of electrical current flow took center stage in electrochemistry. Tafel introduced a relationship between it and the electrode potential, expressed in "Tafel plots," which are graphs showing log of the current versus potential. Butler introduced the concept of active ion formation and derived the Nernst equation in terms of the Tafel relation. These kinetic observations were interpreted with Arrhenius rate theory. Predicting the parameters in these relationships has been one of the goals of theorists ever since. Gibbs had already worked out the formulas that could be used to define states, and in principle equilibrium properties could be determined if enough were known about atomic and molecular energies and their temperature dependencies. Further, if enough were known about them, activation energies could be deduced to fit the kinetic equations. However, decades would pass before they became available with sufficient accuracy to initiate such studies. In the meantime, thermodynamic equilibrium properties were viewed through the Nernst equation and kinetic properties were viewed through the Tafel equation.

E. Schrödinger's discovery of quantum mechanics in 1926 sparked rapid advancement in the understanding of the properties of matter. In 1931 R. W. Gurney combined the quantum mechanical concepts of (i) band theory and densities of states for electrons in metals (ii) electron tunneling to or from an electrode surface and (iii) molecular densities of states and their occupation according to the Boltzmann distribution, to derive the Tafel relation. The electrode's potential was understood to be tied to its electron chemical potential, or surface Fermi level, and this changed when charge passed into the electrode. In the quantum mechanical based computational methods that evolved when computers became adequate, slabs of atoms several atom layers thick were employed. The Fermi level of these slabs corresponded to electron chemical potential of the electrode surface. Connecting these slab electronic structures with the bulk material of the electrode with its own Fermi energy is made with the band bending and concept from semiconductor physics. The electron charge resided in the surface region of the metal electrode and the electrolyte, containing anions and cations, became polarized to form a stable double layer with no net charge in the interfacial system. Though Gurney's derivation made use of assumptions and approximations, it provided the modern underpinning for thinking about the kinetics behind Tafel plots. Born-Oppenheimer vibrational-rotational energy surfaces and densities of states coming from the Schrödinger equation provided visualization of electron transfer processes between different states of a system and provided parameters for calculating transfer probability. Changes in electrode potential were modeled by moving the Born-Oppenheimer energy surfaces up or down on the energy scale. The surfaces were assumed to be rigid, with their shapes independent from the electrode potential. A value for energy of the oxidized species was assigned by adding the energy of an electron produced from the Fermi level to the energy surface. According to the Franck-Condon principle, the reduced species could be oxidized to this state with an activation energy when the Born-Oppenheimer structures were the same. Since the electron comes from an external source and not the electrode, changing potential did not move the electrode surface from the potential of zero charge (pzc). The expected influence of charging the electrode surface on bonding to the surface, which would modify energetics of the reaction, was absent. Theoretical analyses were based on this model until the 1990s when computers with power sufficient to model the electrochemical interface by adding electron charge to the surface entered the scene. Nevertheless, this picture, which traces back a century, continues to provide a widely used model for electron transfer.

A model was built with Born-Oppenheimer potential energy functions for the reduction of $H_3O^+$ to generate H atoms. In 1936 Butler introduced a bond between the H atom and metal atom M of the electrode surface and thereby extended what was called the "outer-sphere" model, to "inner-sphere" reactions, meaning part of the solvation shell was removed to allow bonding to the electrode. With the M–H bond potential energy added, the activation energy for reduction of $H_3O^+$ was lowered, and since metal surfaces could evolve $H_2$ this lowering was possibly first example of electrocatalysis.

Up through the 1960s computations employing changes in electrode potentials continued using the rigid Born-Oppenheimer potential model. The energies of the ions in the electrolyte were components of the Gibbs energy of reactants and products during oxidation and reduction reactions at the electrode. The potential of a half-cell was a function of the Gibbs reaction energy and was expressed relative to the value zero for the standard hydrogen half-cell. The Born model for predicting solvation energies of the ions was introduced in 1920. In it the ions were modeled as charged conducting spheres which interacted, according to the equations of classical electrostatics, with a dielectric medium representing the electrolyte. In 1939 Latimer, Pitzer and Slansky parameterized the Born model with Pauling's ionic radii of 1927, increased by 0.1 Å for anion and 0.85 Å for cation sphere sizes, and calculated solvation energies in remarkable agreement with measured values.

In 1952 Libby proposed a model for electron exchange between ions in aqueous solution that used rigid Born-Oppenheimer potentials but Marcus pointed out that the model violated energy conservation. In 1956 Marcus, beginning with the Born model, developed what is known as the Marcus theory in a series of papers up to 1965 for electron transfer between solvated ions and for electron transfer reactions at electrode surfaces. In 1959 George and Griffith introduced hydration shell breathing mode vibrational energies for the ions which when, added to the Libby model, corrected it for predicting electron transfer activation energies. Marcus incorporated breathing mode energies in his model, and this completed his theory, whose main predictions were confirmed experimentally by Closs et al. in 1986. Experimental confirmation led the Nobel Committee to award Marcus a Nobel Prize in 1992. The Marcus theory, which used Gibbs energy surfaces rather than Born-Oppenheimer potential energy surfaces, continued to model electrode potential changes by shifting the surfaces on the energy scale. Until semiempirical, ab initio, and first principle density functional quantum chemical modeling calculations began to take off in the 1980s, most researchers explored pzc models mathematically to explain potential, activation, and current flow within the Butler-Volmer and Marcus frameworks.

In the 1970s and 1980s high vacuum chambers became important resources for spectroscopic determination of surface structures and following reactions taking place on surfaces in vacuum. Several evolving surface sensitive spectroscopies led to increasing analytical and structural understanding for the surface-vacuum interfaces. Ordered single crystal surfaces in vacuum were exposed to controlled amounts of adsorbing molecules and structures, and electronic and vibrational properties were extracted from the measurements.

Several of the spectroscopies could be used to obtain in situ vibrational, electronic, and structural information for electrode-electrolyte interfaces as well. From these spectroscopies two important features of electrochemical interfaces were discovered. First, changes in the work function, which is the negative of the Fermi energy of an electrode surface, were found proportional to changes in the applied potential. Second, bonds of adsorbed molecules were found to be robust in the double layer environment. Shifts in Fermi energy affected the bonding of adsorbate molecules to electrode surfaces and changing the applied potential perturbed their vibrational frequencies and electronic energy levels by spectroscopically measurable amounts. The measured shifts could be explained and qualitatively predicted using semiempirical perturbation molecular orbital theory. An alternative viewpoint used the dependence of approximate double layer electric fields on electrode potential. In this case the Hamiltonian for a molecule, either adsorbed on a surface or detached, had added to it the electric field as a perturbation and the Schrödinger equation was used to calculate the field-dependent properties of the molecule. In both models chemical bonds could be tracked, undergoing softening and then breaking and forming new bonds as the potentials or fields corresponding to the reversible potential were approached.

The new spectroscopies provided data by which the emerging quantum calculation codes and the models used for interface structures could be tested. In the 1980s and 1990s researchers calculated trends for potential dependencies of vibrations and reactions of adsorbed molecules by applying semiempirical molecular orbital theory, where the hamiltinian used in the Schrödinger equation omits electron repulsions and some matrix elements are set to zero. By changing parameters of the theory to shift the Fermi energy proportional to changes in electrode potential, they found it possible to calculate trends that agreed with measured vibrational frequencies and molecular orbital energy levels in photoelectron spectra. Meaningful treatments of electron transfers and electrocatalytic reactions were, however, beyond the capabilities of the semiempirical models.

Any theoretical breakthrough required greater computing power, and such power was increasing rapidly. In the 1990s and 2000s ab initio Hartree-Fock (HF) and configuration-interaction (CI) and first principles density functional theory (DFT) could finally be applied to electrode surface models to study reactions. The electron wavefunctions from these approximate applications of the Schrödinger equation were self-consistent, meaning each electron experienced the coulomb influence of all the other electrons as the system energies were minimized. Self-consistent field (SCF) theories were advancements over the semi empirical codes where wavefunctions and molecular orbital energy levels were dependent on atomic orbitals and Hamiltonian matrix elements derived from parameters and rules. With the electrode surface represented by a cluster of atoms, self-consistent calculations of electron transfer reaction energies yielded activation energies and reversible potentials of relatively dependable accuracy. Like the work of Butler and others mentioned earlier, in the early applications of SCF codes the electron source or sink remained remote. Transition states for electron transfer reactions were found by searching for structures where the electron affinity (EA) or ionization potential (IP) of the reaction center matched the external electron source or sink potential.

Applications of the SCF methods became largely directed toward energy and environmental applications, including hydrogen and methanol fuel cell activities and reaction mechanisms, as well as water electrolysis, and fuel cell catalyst optimization. Research in these areas initiated largely in response to interest from industry and the government. Major funding in the US came from various companies, the Department of Energy, the Army and the National Science Foundation.

The electronic structure of the hydrogen fuel cell is introduced. The chemical equilibria at the anode and cathode surfaces and the double layers formed before connecting the electrodes with a wire and proton-conducting electrolyte are depicted. A sufficiently catalytic material such as Pt establishes the equilibria quickly. After the electrodes are connected, electrons are injected into the Pt conduction band and then a connecting wire and will do work on a resistive load on the way to the cathode. Additional experimental proof of the electron chemical potential of electrons in solution having physical significance is provided by experiments where electron holes ($h^{+}_{dia}$) form on H-terminated diamond surfaces exposed to the ambient. These make the diamond surface into a p-type conductor. This also happens for oxides and recent literature shows transistors can be made using this surface phenomenon.

The electron chemical potential depends on pH and this makes it possible to for insulators with appropriate bands in the gap to have pH-dependent occupations and optical properties.

Understanding electrocatalytic reduction of $O_2$ and catalyst optimization was an active topic for experimental researchers and, as methodologies improved, began drawing the interest of advanced theoretical research groups as well. The earliest self-consistent ab initio and DFT studies modeled the surface reactions as if they were at the vacuum interface. These calculations omitted the electron transfers and potential dependencies when calculating reaction mechanisms and activation energies. The results were of uncertain relationship to the electrochemistry. Attention turned to calculating reversible potentials for electron transfer steps. The reduction of $O_2$ at the cathode in a fuel cell involves four electron and proton transfers to the intermediates and each has a different standard reversible potential in solution, spanning a range of 2.84 V. The goal in oxygen reduction catalyst development is to bring them all to the same and optimal value, 1.23 V for standard conditions or the appropriate value determined by the Nernst equation for the electrolyte being employed. But this is not enough. To be successful, the electrocatalyst must also present low activation energies for the electron transfers. The reversible potential for the electron transfer reaction could be predicted, perhaps better to say estimated, from the molecular reaction energy plus the electron energy that put the electron transfer at equilibrium. One widely used method today employs the computational hydrogen electrode (CHE) model, which uses the work function of the standard hydrogen electrode for reference. Another method is the linear Gibbs energy relationship (LGER) model, which uses solution phase standard reversible potentials as reference. The CHE model is especially popular and has been used in hundreds of publications. Calculated reaction energies for the four steps, when graphed according to the sequence of electron transfers, provides visualization of spontaneity and identification of energy bottlenecks. The LGER method directly produces reversible potentials for the electron transfer steps and identifies high overpotential bottleneck steps. These procedures, like previous theories, still did not treat all of the effects of the applied potential since the models were fixed at the pzc and therefore omitted electrode surface charging and electrolyte polarization.

A self-consistent DFT code, written during 2006–2008 by R. Jinnouchi, included surface charging and compensated it with a polarizable ionic electrolyte. This made possible calculations of reaction energetics on electrode surfaces as functions of potential. With this code the yoke of rigid potential-independent Born-Oppenheimer potentials that were the basis of past models was removed. More accurate predictions of reversible potentials for bulk and surface electron transfer potentials became available and research groups worldwide began developing similar procedures, greatly expanding accurate theoretical research into electrocatalysis. Increasing computational power means more topics will be explored and increasingly better predictions will ensue in future with these codes. With addition of artificial intelligence (AI) and machine learning (ML) and growing computer power it is now probable that far more accurate predictions of mechanisms and energetics will be achieved. Development of more efficient energy conversions and electrochemical syntheses of molecules and advanced materials may be expected. It is possible that in future AI will be viewed as having caused a scientific revolution.

Cleveland, USA Alfred Anderson

**Acknowledgements** First, I am happy to acknowledge my primary mentors in science. Robert G. Parr with whom I did my graduate work at Johns Hopkins University instilled in me an attitude of precision in my scientific thoughts and writing. The two years I spent in Roald Hoffmann's lab at Cornell University as a postdoctoral associate taught me Molecular Orbital Theory by the examples he presented in the fascinating group meetings. At the time he was changing his focus from organic to inorganic chemistry and we in the group all benefited as we, like understudies following their troupe leader, entered this fascinating new domain with him. The camaraderie between members of the group also contributed to making this a wonderful experience. Individuals included Jeff Hayes, Steve Kim, Mihai Elian, Jack Thibeault, Bob Bissell, Angelo Rossi, Hiroshi Fujimoto, Notker Rösch, Nguyên Trong Anh, and Maynard Chen. At the time Roald was getting interested in surface chemistry and he sent me to group meetings run by Thor Rhodin in the Cornell Physics Department. Surfaces subsequently became a focus of my research. I also benefitted from time spent with Harrison Shull at Indiana University, Arthur Hubbard at the University of California at Santa Barbara, and as a J. Willard Gibbs Instructor in the Chemistry Department at Yale University. When I joined the faculty at Case Western Reserve, Ernest Yeager set for me the goal of understanding oxygen reduction electrocatalysis and I have tried to justify confidence in me.

At Case the delightful interactions with my students, postdoctoral researchers, Case colleagues, and many others helped keep my interest in science and spirit of inquiry alive. Funding, acquired through various industrial sources and federal granting agencies, was often goal oriented, perfect for an applied theorist like me using easily applied semiempirical models. There was sufficient freedom to advance the modeling and apply new perspectives to entrenched problems. Inspiring collaborations were had with established scientists off campus including Bob Grasselli, Bob Swofford, Jim Burington, Mike Ward, and Jim Brazdil at The Standard Oil Company (Ohio), and Jim Smialek at NASA Lewis Research Center, Phil Ross at Lawrence Berkley Laboratory, Kieth Hall at the University of Pittsburgh, Jurgen Kreuzer at Dalhousie University, Branco Popov at the University of South Carolina, Craig Hill at Emory University, Sanjeev Mukerjee at Northeastern University, and Ed Holby at Los Alamos National Laboratory.

Interactions with campus colleagues include collaborations with Ernest Yeager, Fred Urbach, Malcom Kenney, Wiley Youngs, John Fackler, and Dan Scherson in the Case Chemistry Department; Arthur Heuer in Materials Science; and John Angus, Uziel Landau, Heidi Martin, and Rohan Alolkar in Chemical Engineering.

High school students contributing in my lab included Kathy Collen and Kelly Acker.

Undergraduate Terry Gordon contributed.

Graduate students included George Fitzgerald, Dave Nickles, Barbara Trzcinska, Dae Bok Kang, Mike McDevitt, Robin Grimes, Mohamed Awad, Dave Ewing, Sung Y. Hong, Paul Shiller, Jenwei Yu, Shu Fen Jen, Titus Albu, Yu Cai, Tianhou Zhang, Kiera Kurak, Haleema Asiri, and Meng Zhao.

Postdoctoral associates spending time in my lab include Naba Ray, N. C. Debnath, Yunsoo Kim, S.-C. Hung, Satya Mehandru, David Onwood, C. Ravimohan, John Maloney, Kashi Nath, Paul Shiller, Seeyearl Seong, Rey Sidik, Jay Narayanasamy, Jerome Roques, Hans Schweiger, Elen Vayner, and Jamal Uddin, Sang Joon Choe.

Scientists who visited my lab for extended periods include Ryosuke Jinnouchi from Toyota Central Research, Jeff Nichols from Malone College, Zeki Al-Saigh from the University of Charleston, San Yan Chu from National Tsing Hua University, Munime Lundeen from Marmara University, Emilia Grantscharova, Lubomir Kostadinov, and Nikolay Neshev from the Bulgarian Academy of Sciences, Donald Dowd from Allegheny College.

This book, by focusing on the development of theoretical models and applications in electrochemistry, is able to highlight the efforts of several of my collaborations over the years which added to the interest and understanding that made the book possible. To everybody, thank you!

As a final note, I thank Case Western Reserve University and the Chemistry Department for providing office space, computer and library facilities that were used during the writing of the book. In addition, the free parking available to emeritus faculty was helpful.

**Competing Interests** The author has no competing interests to declare that are relevant to the content of this manuscript.

# Contents

# Chapter 1
# Classical Theories for Electrochemical Measurements and Electrocatalysis

**Abstract** This chapter highlights the development of theories for interpreting electrochemical measurements which span over one hundred years. It began when Volta and Nicholson constructed batteries and Davy used batteries to reduce metals and generate gases. Davy developed an electrochemical series making use of the periodic table of elements. Faraday gave further structure to the observations. Thompson showed electricity is made of electrons. Nernst demonstrated a relationship between electrolyte concentrations and cell potentials and Gibbs developed the formulas of thermodynamics and explained electrical work in terms of charge passed from one potential to another. Tafel, Butler, and Volmer adopted current flow measurements to analytic formulas. Schrödinger's wave equation provided the understanding of electron behavior which was used by Gurney to explain electron transfer at electrode surfaces. The analytic formulas cell potentials from concentrations and for current flow from potentials remain foundational to electrochemical science and engineering.

## 1.1 General Introduction

This chapter briefly introduces the developments in chemistry, physics, and electrochemical measurements and interpretation that are the foundational starting points for the theories for electrochemistry presented in later chapters.

## 1.2 Volta Through Faraday: Phenomena and Definitions in Electrochemistry

Discovery of electrochemical phenomena began around 1800 when Count Alessandro Giuseppe Antonio Anastasio Volta constructed a zinc-copper battery using spacers between the metal electrodes wet with brine [1]. Such batteries were called voltaic or galvanic piles. Having learned of this, the same year William Nicholson and Sir Anthony Carlisle constructed their own voltaic pile and used



A. Anderson, *Past and Current Models in Electrochemistry*,
https://doi.org/10.1007/978-3-031-97791-6_1

it to decompose water by passing a voltaic current [2]. Sir Humphry Davy in the first decade of the century used electricity from batteries to prepare alkali metals, alkaline earth metals, and halogen gasses and he proposed an early electrochemical series [3]. The electrical oxidation and reduction of metals continues as the key processes in battery and metal plating and purification technologies today.

Michael Faraday established much of the language used in electrochemical science and first presented it in his 1834 paper [4]. Faraday, a tireless experimenter, possessed clarity of logic that allowed him to see correctly the implications of measured data. He avoided having his thoughts restricted by theoretical dogmas. This attitude was expressed, for example, in his "Lecture on the Forms of Matter" to the City Philosophical Society in London in 1819: "All theories are fixed upon uncertainties and all of them want alteration and support" [5]. Volumes I and II of "The Life and Letters of Faraday" [5] put the reader in close intimacy with Faraday's personal experiences and philosophical (now called scientific) thoughts. These letters are wonderfully written, emotionally and intellectually moving, and should be read by curious scientists as well as non-scientists who would like to know how the great scientist thought.

## 1.3 J. J. Thompson and the Electron

In the early- and mid-nineteenth century, the theoretical understanding that we possess today was hardly imaginable, but toward the end of the century, things began to percolate. Faraday discovered the concept of electrical and chemical equivalents, including the relationship between the amount of electricity and the amount of reaction–one Faraday is 96,485 Coulombs where one coulomb is the amount of charge passed when an electrical current of one ampere flows for one second. Since the discovery of the electron in 1897 by J. J. Thompson, who received the Nobel prize in physics in 1906 [6] we now know a coulomb is the equivalent of $6.242 \times 10^{18}$ e, where e stands for the charge of one electron, and that a Faraday of electron charge contains $6.022\ 10^{23}$ electrons. The word 'electron' was the invention of George Johnstone Stoney in 1891 [7]. Stoney had earlier introduced the concept "atom of electricity".

## 1.4 Nernst and Gibbs: Cell Potentials and Concentrations

The Nernst equation, relating equilibrium potentials to ionic concentrations in cell electrolytes, appeared in 1889 [8]. Nernst also discovered the third law of thermodynamics which at the time was known as the new heat theorem [9].

J. Willard Gibbs incorporated thermodynamics into electrochemistry in papers published in the Transactions of the Connecticut Academy III from1876 to1878 [10]. He showed the available energy, $\zeta$, (p. 87), now called the Gibbs energy and symbolized by $G$) of a reaction when reversibly generating electrical work is proportional to the potential difference of the two electrodes and the amount of electron charge passed (p. 394). He arrived at formula for potentials more general than the Nernst equation. Gibbs also laid down the science of statistical mechanics [11] In this work, he defined the density-in-phase function, D, (p. 6), now called the density of states function.

At this time, the whole field of modern equilibrium electrochemical science was ready to launch, with its understanding based in thermodynamics of molecules, and ions and electrons, the particles making up electricity. In terms of densities of states, properties of matter became understandable and predictable, but the accurate determination of densities of states awaited the discovery of the Schrödinger equation and the quantization it correctly predicted. But before this, let us look at what was known about kinetics in electrochemistry.

## 1.5 Tafel, Butler, Volmer: Kinetics of Current Flow

While the Nernst equation provided successful predictions of the concentration dependencies of cell reversible potentials (for low enough concentrations), it said nothing about reaction rates. They required another theory. Tafel measured rates of reactions at fixed potentials, and discovered a relationship between polarization, $e$ (now usually called the overpotential, $\eta$), and the current density $I$ (current per unit electrode area, now usually called $i$) for hydrogen evolution in acid [12]:

$$\boldsymbol{e = a + b \log I} \quad \text{or } \eta = a + b \log i \tag{1.1}$$

This is the famous Tafel equation, in which $a$ and $b$ are empirical parameters, differing for electrodes of different metals. For $H_2$ evolution Tafel proposed the mechanism

$$2H^{+}(aq) + 2e^{-} \rightarrow 2\,H(ads) \rightarrow H_2(g) \tag{1.2}$$

In 1924, Butler initiated the concept of activated ion formation at metal electrode surfaces [13] and by equating the formation and deposition rates, the Nernst equation was derived. This also applied to reversible potentials of ions over inert electrodes [14]. Butler found the formalism consistent with the observed linear dependence of overvoltage for hydrogen oxidation on log of the current density in the low overpotential region [15]. Measurements by Erdey-Grúz and Volmer supported the formalism too [16]. They additionally found it applied when depositing several metals in the low and high overpotential ranges [17].

## 1.6 Schrödinger, Sommerfeld, and Gurney: Band Theory and Electron Transfer

In 1926 E. Schrödinger introduced his quantum wave equation [18]. Application to charge transfer at the electrochemical interface followed after several years passed. According to the belief prior to the development of quantum mechanics, the picture of electron transfer was simply that it occurred when, for example, a positive ion collided with the electrode surface. Details of what happened during the collision were unknown. But in 1931 R. W. Gurney applied the newly discovered quantum mechanics to electrolysis [19]. Using the concept now called tunneling, Gurney softened the collision interaction, allowing the possibility of electron transfer occurring over a range of distances between the ion and the electrode surface. In his formalism, the probability for electron transfer depended on the probability of energy-matching of electron states in the electrode surface and electron states in the molecule or ion participating in the electron transfer, satisfying energy conservation, multiplied by the probability for electron tunneling. At the time it was too difficult to use wavefunctions to evaluate the general formula for electron transfer. Gurney therefore assumed the molecules involved in the electron transfer were all at the same mean distance from the electrode surface, which is the simple classical picture with parameterization of the quantum mechanical probabilities. Sommerfeld's free-electron band theory provided the electron energies [20, 21].

Gurney set out to find the relationship between the log of the current density, $i$, and the electrode potential and its dependence on temperature. He presented the following model, expressed here using modern notation. It was developed apply to radiationless electron transfers, meaning, for example, that during an electrochemical reduction reaction the energy of the electron in the electrode was the same after it reduces a molecule or ion. That is, no energy was lost in the form of radiation by photon emission or by vibrational excitation during the transfer. Gurney proposed electron transfer probability could be described by the product of three terms. They are: (i) the probability for an electron incident on the electrode surface to have energy $E$, related to the applied potential $U$, (Gurney used the symbol V) in the cathode, $n(E,U)$. (ii) the probability for tunneling through a potential barrier of width $x$, $P(E,x)$. (iii) the probability that a positive ion has an electron affinity energy equal to $-E$ when incident on the other side of the barrier at position $x$, $N(E)$. The last is the condition of resonance between the electrode as an electron donor and the electron acceptor. The electron accepting species need not be positive and may or may not be charged. Using $n$ and $N$ normalized to unit area at the electrode-solution interface, the current densities are proportional to the overall probability. The acceptor ion energies $E$ sometimes stood for Gibbs free energies but internal energies were often used in their place, and they included vibrational–rotational energies.

## 1.7 Gurney's Model for Electrode Current Densities and Role of Fermi, Dirac, Born, Oppenheimer, and Boltzmann Formalisms, and the Franck–Condon Theory

In the modern view, increasing potential means higher workfunction. This relationship stems from Sommerfield's band theory wherein charge added to a metal in vacuum essentially resides on its surface and changes the surface ionization potential or electron affinity. It was known that the work of ionization is the Gibbs free energy for moving a charge through a potential. Many years passed before changes in electrode workfunction and changes in the applied potential at the electrochemical interface were experimentally observed to be proportional. This is reserved for later discussion.

Gurney used the inverted relationship of potential and workfunction in his papers and had potential decreasing as workfunction increased. In the following discussion, $E$ stands for internal energy or Gibbs energy, as appropriate to context, for sometimes internal energy is employed as an approximation to the Gibbs energy. Using the free-electron model for metals, the electron distribution function, $n(E,U)$, was the Fermi–Dirac distribution over free-electron states at temperature $T$ [22]. This was assumed to be constant over small energy ranges, and the distribution function of the negative of the electron affinities, $EA$, for the species undergoing reduction, $N(E,x)$, with $E = -EA$, was a Boltzmann distribution over vibrational–rotational states, assumed to be continuous at temperature $T$. The energies of these states were taken as quantum mechanical solutions of the Born–Oppenheimer potential energy functions using the vibrational–rotational Hamiltonians [23]. The hydration energy of the solution species entered parametrically but not the interaction energy of adsorption on the electrode surface. To avoid the need for molecule or ion wavefunctions, Gurney assumed all electron transfers take place for the same distance, $x$. With these approximations, the transfer probability integral

$$\iint n(E, U)P(E, x)N(E)dEdx \tag{1.3}$$

was easily approximated. The measured current density is proportional to this integral.

The factor $n(E,U)$ in Eq. (1.3) is proportional to $|E|^{1/2}\exp\{[-E-\varphi_{pzc}-e(U)]/kT\}$ where $\varphi_{pzc}$ is the thermodynamic workfunction for the electrode at the potential of zero charge (pzc), $U_{pzc}$, (Gurney used the symbol $E_1$ for the electrode with no potential applied) and $e(U)$ is the shift from $U_{pzc}$ caused by applying a potential. The term $-\varphi_{pzc}$ is also the electron chemical potential, $\mu_e$, for the uncharged surface, and it plays the role of the Fermi energy or Fermi level ($E_f$), when a thin slab of several atoms thickness represents the electrode surface. In later chapters more will be said about this, and in particular how this thin slab interfaces with the bulk electrode material which has generally a different Fermi energy and how electrons make it through an external circuit between the anode and cathode. Applied potentials shift $E_f$ by $-e(U)$. This expression for $n(E,U)$ applies for electron energy values that are greater than $E_f$. The density of these states is sufficient for theoretical treatment to high current densities. These days the standard hydrogen electrode is the reference for potential and it is connected to the vacuum scale by its workfunction, approximately, 4.43 eV, at 0 V (SHE).

For energy levels below the vacuum level, the energy $E$ has negative values. It was at the time not possible to assign accurate values to the Fermi energy on the SHE scale. The ion Boltzmann distribution function was $N(E) = \exp[(E-E_o)/kT]$, and the electron tunneling factor was $P(E,x) = \exp\{-4\pi x'/[h(-2\ m|E|)^{1/2}]\}$ where $x'$ is the mean distance between ions and the electrode surface. Using the modern convention for depicting energy levels, Fig. 1.1 illustrates the distribution functions and the effect of decreasing the electrode potential from $U_{pzc}$ to $U$, modeled by shifting the Sommerfeld band up. It is evident that if the potential is high, as depicted by $U_{pzc}$, there is no overlap and no electron transfer. However, reducing the potential shown in the left-hand set of energy levels in Fig. 1.1 so that electron and ion distribution functions begin to overlap makes electron transfer possible, as shown by the right-hand set of energy of levels. Chapter 6 takes the viewpoint that the Fermi level of the surface shifts to maintain equilibrium with the polarized electrolyte but the band levels in the bulk do not so that the connection between the surface and bulk phases may illustrated as a rapid band bending.

Gurney's prototype reaction was $H_3O^+$ reduction to $H-H_2O$ as a model for the first step in hydrogen evolution over metal electrodes. In his first application of Eq. (1.3) he took the energy in the Boltzmann distribution to depend on the excitation energies, $E_n^*$, in the rotational-vibrational spectrum of the cation, as shown in Fig. 1.2. Integration over the small range of $E$ from $E_0$ to $-\varphi - eU$, for which the integrand contributes significantly, yielded the equation for the log of the current density:

$$\log i = \left(E_0 - \varphi_{pzc} - eU\right)/(kT) + \text{additional terms} \tag{1.4}$$

For $H_3O^+$ reduction, $E_o$ was estimated to be about 4 eV. This is based on a 13.6 eV hydrogen atom ionization potential (quoted by Gurney as 13.5 eV) and on an assumed hydration energy for the proton of about –9 eV. Gurney showed that the additional terms make small contributions.

It was assumed the proton attached to a water molecule forming stable $H_3O^+$ (hydronium ion), with strong O–H bonds and short internuclear distances. By contrast, the H atom was assumed to be repelled by the O atom in $H_2O$ and the distance-dependent repulsion energy was given as $R(r)$. In Fig. 1.2 the hydration energy of $H_3O^+$ is neglected and quantized energy levels are shown. However, as Gurney stated, variations in local environment and structure effectively smeared them out creating a near continuum of energies for the purpose of integration. Figure 1.2 the zero for the energy scale is assigned zero for the isolated bare proton, the isolated water molecule, and the electron with zero energy at the vacuum level (Gurney used for the zero of energy the isolated bare hydrogen atom and isolated water molecule).

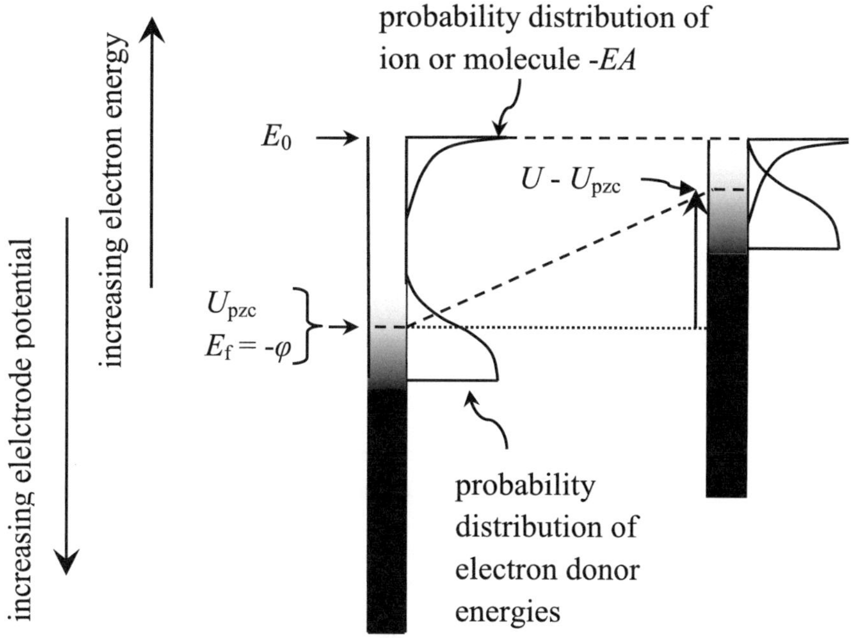


**Fig. 1.1** The metal valence band shown in the left column qualitatively shows the Fermi energy, the probability distribution of electron donor energies around $E_f$, and the higher energy probability Boltzmann distribution function of acceptor vibrational states in the species to be reduced. Arrows indicate how shifting up of the valence band happens when the electrode band energy increases as the potential is decreased. The column on the right shows how when the metal band moves up due to decreasing the potential, the electron distribution functions of the surface overlap with the Boltzmann vibrational distribution allows resonant electron transfer

The curves show the structure dependencies for the $H^+ + H_2O$ vibrational potential curve, $W(r)$, and the $H + H_2O$ repulsion energy, $R(r)$. A single $H_2O$ molecule acts as a surrogate for the bulk water so that as the H–O internuclear distance increases the proton is leaving the surface of the water phase and entering vacuum. The energy $E$ that appears in the equations above and in Fig. 1.2 is equal to $W(r)$–$R(r)$ is positive which means the activation energy is for oxidation of $H_3O$.

When the hydration energy of $H_3O^+$ is included, Fig. 1.2 is changed because the $H_3O^+$ potential curve is stabilized because the proton hydration energy is about 11.4 eV, using a contemporary estimate by Zhan and Dixon [24]. According to Zhan and Dixon, hydration also stabilizes the electron by about 1.5 eV, but this is not relevant to the Gurney picture where the energy of the electron that is available for the reaction at the surface is determined by the Fermi energy of the electrode, that is, by the applied potential. Consider the equilibrium in Eq. (1.5).

$$H_3O^+ + e^-(U) \rightleftharpoons H_3O \tag{1.5}$$

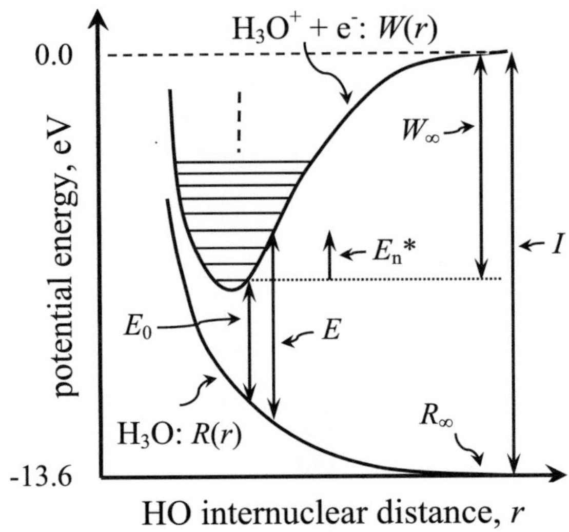


**Fig. 1.2** Qualitative graph of potential energies, $E$, as functions of HO internuclear distances, $r$, in $H_3O^+(g) + e^-$ (upper bonding energy curve) and $H_3O(g)$ (lower repulsive energy curve). At the dissociation limit $H^+(g) + H_2O(g)$ is 13.6 eV less stable than $H(g) + H_2O(g)$, and the difference is the ionization potential, $I$, of H(g) or the negative of the electron affinity of $H_3O^+(g)$. At intermediate values of $r$, $E$ is ionization potential of $H_3O(g)$ and is also the negative of the electron affinity of $H_3O^+(g)$; it is the same as the energy appearing in Eq. (1.3) of the text. Also shown are $E_0$, which equals the energy of the ground vibrational state of $W(r)$–$R$(r), and $E$, which equals $W(r)$ in the nth vibrational energy level—$R$(r)

When $H_3O^+$ and $H_3O$ have the same structure and the potential has a value corresponding to the $E_0$ transition in Fig. 1.2, then if the potential is increased, meaning $E_f$ moves down on the energy scale then, to maintain equilibrium for the structure, the $H_3O^+$ potential curve must be stabilized so that energy for the reaction is again conserved. Therefore, the $H_3O^+$ is moved down or up by the same amount as $E_f$ moves down or up. Hydronium ion hydration actually causes Born–Oppenheimer energy curves in Fig. 1.2 to cross and the consequences are discussed a bit later. It is convenient for now to keep them separated as in Fig. 1.2, as Gurney did in his presentation.

Electron transfers are assumed to be rapid, in accordance with the Franck–Condon principle, meaning they take place before the vibrating atom nuclei move distances that would complicate the model [25]. Figure 1.2, shows that for thermally occupied higher energy states the ionization potential of $H_3O$, $E_n$, is greater than $E_o$. In the same paper, Gurney introduced and discussed an analogous figure for neutralization of a negative ion. Looking at Fig. 1.2, one can see that $E$ increases at a rate faster than the rate of increase in vibrational–rotational energies. Examining in detail, one can define two angles, as shown in Fig. 1.3. Approximating the upper and lower curves as tangential lines, an angle for the upper curve, $\Theta_W$ and for the lower curve, $\Theta_R$ can be defined. Using these angles and the energies in the figure, one can derive the

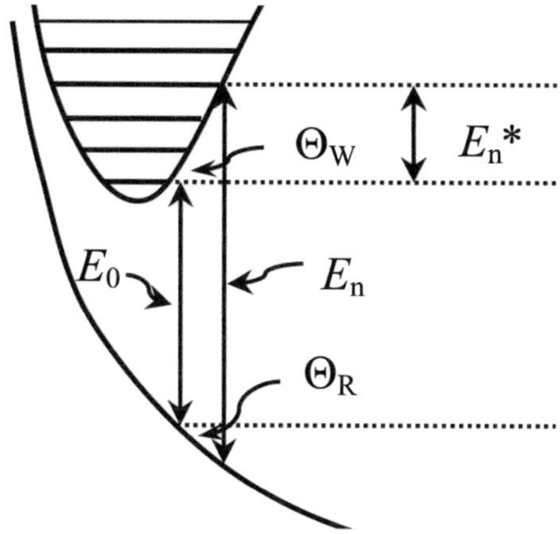


**Fig. 1.3** Detail of Fig. 1.2, defining tangential angles $\Theta_W$ and $\Theta_R$. $E_n = W(r)$ in the nth vibrational state—$R(r)$, $E_0$ is as before and $E_n$* is $E_n$–$W(r)$ in the ground vibrational state

equation:

$$E_n \approx E_0 + E_n * [1 + |(\tan \Theta_R)/(\tan \Theta_W)|]$$

$$\approx E_0 + \gamma E_n* \quad (1.6)$$

where $E_n^* = E_n – E_0$. When the slopes have the same absolute value, $\gamma = 2$ and the rate of increase of $E$ is double the rate of increase of vibrational energy, $E_n$*. The inverse of $\gamma$ is called the symmetry factor. In Fig. 1.2 $E$ has quantized values $E_n$ but current measurements displayed in Tafel plots, which are graphs of overpotential vs log of current density, are smooth and do not show quantization. This means the density of vibrational–rotational states for the hydrated proton is continuous as far as this measurement is concerned. However, it is helpful to maintain the subscript in the following.

Gurney found that processing Eq. (1.6) in Eq. (1.3) by treating the $P(E, x)$ factor as a constant, and subsequently dropping a small term from the integration, led to the result

$$\log i = \left(E_0 - \varphi_{\mathrm{pzc}} - eU\right)\gamma/(kT) + \log T + \text{constant} \quad (1.7)$$

This has the form of Eq. (1.1), which is the Tafel equation. If the electrode potential is too positive for the reduction reaction to occur, but is decreased until the distributions begin to overlap, the integral in Eq. (1.3) increases exponentially from zero. Figure 1.2, which is based on Gurney's own depiction, did not include perturbation to the lower potential curve caused by forming the metal-H bond; the curve in the figure represents solely the $H–OH_2$ repulsion.

## 1.8 Butler Added Interactions with the Surface to Gurney's Model: The Electrode as a Catalyst

Butler continued with the Gurney model in a paper addressing the overvoltage for hydrogen evolution. His contribution to Gurney's model was the inclusion of stabilization of the $H_3O$ system caused by the chemical bonding interaction between the H and the electrode surface atoms [26]. Butler changed scales in the figures from Gurney's and combined energy terms. A clear way of showing Butler's contribution is to work in the framework of Gurney's approach by making a modification to Fig. 1.2. Butler added an estimated attractive Born–Oppenheimer potential energy curve for an H atom interacting with then metal electrode surface to the estimated repulsive potential curve for H interacting with $H_2O$. Figure 1.4 shows how the potentials add up, along the dashed line, with the result that forming the bond between the hydrogen atom and the metal increases the electron affinity, $E$. The hydronium ion curve will rise going to the right because of interactions with M but this is not shown. The model leaves out what we know in the modern perspective, such as the dependence of the surface-H bond strength varies with potential as do structures and energies of other adsorbed molecules and molecules and ions in the double layer region over the electrode surface.

Again, following small changes in $r$, $\Theta_R$ increases by an amount $\Delta$, and the changes in the three potential energy curves are nearly linear so that it can be shown that.

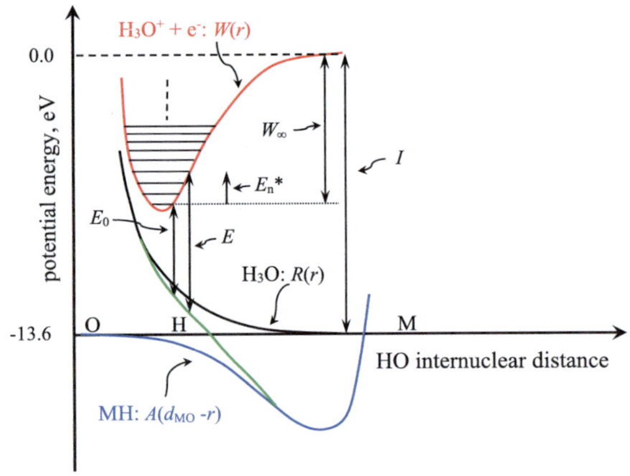


**Fig. 1.4** The Gurney-Butler model. As in Fig. 1.2 but including the MH bonding potential curve $A(d_{MO}\text{-}r)$ (blue) which when added to the repulsive H–O curve (green) to obtain the vibrational potential curve for the O–H–M system (black). The $H_3O^+$ curve $W(r)$ is now red. Red curve to Green curve vertical distances are increased from Fig. 1.2 to give new $E_0$ and $E$ values when O, H, and M are on a linear axis and the distance between the O and M nuclei has a constant value. Reproduced from "How Computational Theorists Deal with Electrode Potential," Alfred B. Anderson 2025 J. Electrochem. Soc. 172 0501 © 2025 Alfred B. Anderson. Published on behalf of The Electrochemical Society by IOP Publishing Limited

$$E_n \approx E_0 + E_n * [1 + |(\tan(\Theta_R + \Delta)/(\tan \Theta_W)|$$

$$\approx E_0 + \gamma E_n^* \quad (1.8)$$

While Eq. (1.8) has the same form as Eq. (1.6), adding the MH potential increases $\gamma$. Equation (1.8) can be interpreted to mean that as the OH internuclear distance increases at fixed MO internuclear distance, so also does the electron affinity of the $H_3O^+$increase, so that the maximum potential of an electrode that can perform the reduction increases. The difference between Gurney's model, where free H atoms formed, and Butler's model, where H formed adsorbed on the metal surface, is that since $\gamma$ is smaller in the Gurney case the electron affinity of the $H_3O^+$ is less for a given MO internuclear distance and the reduction potential for forming H atoms is therefore lower.

It can now be appreciated that as long as two adsorbed H atoms on the electrode surface are able to combine and desorb as $H_2$, a less negative potential and less energetic electrons will be required compared to the forming of $H_2$ by combining two non-adsorbed H atoms generated by $H_3O^+$ reduction. *Thus, an understanding of the electrode as an electrocatalyst for Faradaic generation of hydrogen is drawn. Stabilizing the reduction intermediates activates the overall reduction reaction to initiate at higher potential.*

Butler recognized that the electron transfer would be reversible when the activation energies are the same for the forward and reverse reactions. This is seen by adding the repulsive H-$OH_2$ $R(r)$ curve to a Morse-shaped potential curve for the M-H to give the total potential curve after reduction and aligning it with a Morse-shaped potential curve for $H^+OH_2 + e^-$, as shown in Fig. 1.5a for a given MO internuclear distance, $d_{MO}$. The shapes of these potential functions are assumed to not change as applied potential is changed. This figure omits the vibrational energy levels for $H_3O^+$ and the interaction between $H_3O^+$ and M. The activation energies for the forward and reverse reactions are equal when the electrode potential equals the reversible potential, $U_{rev}$. As long as $d_{MO}$ is constant, any lower potential would increase the amount of reduced species, MH, at equilibrium, and any higher potential would favor the oxidized species, $H_3O^+$. During a collision of $H_3O^+$ with the electrode surface, $d_{MO}$ will not have a constant value. Figure 1.5b shows how to reach the conclusion that for a given electrode potential the activation energy decreases if $d_{MO}$ decreases. It is evident that a more energetic collision will lead to a shorter OM distance at the crossing point, which will lower the average activation energy and increase the probability of reduction at the reversible potential. Looking at Fig. 1.5a, this dependence of activation energy also applies for non-equilibrium potentials.

The ideas derived from examining Fig. 1.5a, b are most rigorous over the ranges in internuclear distances where the Born–Oppenheimer vibrational potential energy curves are approximately valid. Including $H_3O^+$ interactions with the surface M would certainly change the shape of the $H_3O^+$ curve as the ion moves closer to M, corresponding to relatively energetic collisions. However, for more gentle thermal

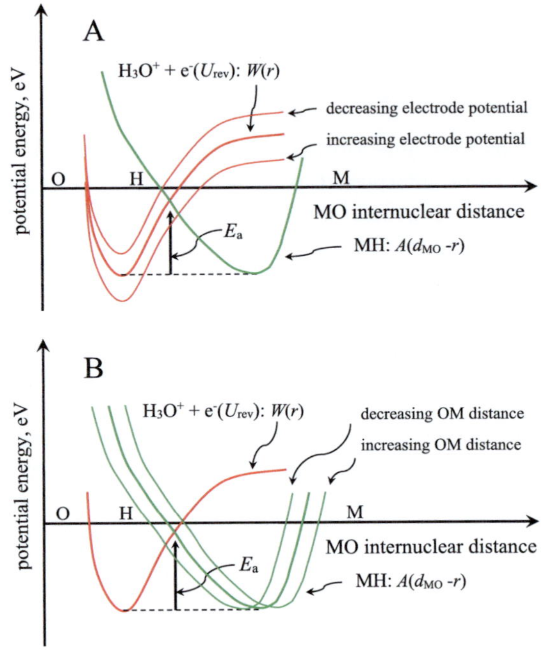


**Fig. 1.5** As in Fig. 1.4 but drawn to illustrate activated reduction near the reversible potential. Panel A shows how when the MO internuclear distance is constant the reversible potential, $U_{rev}$, and the activation energy, $E_a$, may be found by adjusting the $W$(r) curve up and down (changing the potential) so that its minimum matches the minimum in the M-H bonding curve. Panel B shows how when the $W$(r) is fixed (potential constant), decreasing the MO internuclear distance lowers the activation energy. Reproduced from "How Computational Theorists Deal with Electrode Potential," Alfred B. Anderson 2025 J. Electrochem. Soc. 172 036501 © 2025 Alfred B. Anderson. Published on behalf of The Electrochemical Society by IOP Publishing Limited

motions the curves in the overlapping regions shown in the figures are assumed to be rigid enough to support the above interpretations concerning activation energies.

The Tafel equation and Butler-Volmer theory derived from the above elementary model are the basis for modern interpretation of activation energies and current measurements. The model is a success. With the development of powerful computers and advanced programs for solving the Schödinger equation, diatomic two body Born–Oppenheimer potential functions could be replaced by multi-dimensional vibrational energy functions. Such calculations are within reach for treating the reaction model species assumed in Figs. 1.2, 1.3, 1.4 and 1.5 but they are not generally expected to lead to significantly modified results and interpretations. It makes sense to view the Gurney theory and Butler-Volmer model as a zeroth-order descriptions of electron transfer at the electrochemical interface and $H_2$ generation to which perturbative corrections may be added. Important perturbations are the solvation energies of reactants and products, interactions with other molecules on the electrode surface, integrating over all orientations up to the transition state, corrections to the Born–Oppenheimer Hamiltonian for H, which is a light atom, and perhaps most significantly, surface charging and reactant, product, and electrolyte response as the potential is controlled by changing the surface atom charges.

As it turns out, progress addressing the corrections has been slow until the past two decades. Computational theorists are always looking for new challenges but persisting lack of sufficient computing power and a view that electrochemistry is a difficult empirical subject outside their interest have stood in the way. Before introducing the recent theoretical developments based in computational quantum mechanics, there are additional significant developments based on elementary levels of modeling to discuss.

## 1.9 Horiuti and Polanyi: Influence of Electrode Potential and Electric Field on Butler's Model

In 1935 Horiuti and Polanyi deduced a picture equivalent to Fig. 1.5a for M-H oxidation to hydronium ion on a metal electrode. The picture shows the activation energy for oxidation increasing as the electrode potential is deceased [27]. They based their explanation on the assumption that "the full polarization voltage resides between the metal's electronic surface and the layer of adsorbed hydrogen ions, and that it constitutes a homogeneous field." The perspective was that of the $H_2O$–$H^+$ Born–Oppenheimer potential function shifting up or down on the energy scale by the amount $-e\Delta U$ where $\Delta U$ is the change in electrode potential. The result is the same as the Gurney-Butler-Volmer model but it brings double-layer electric fields into the discussion. Since then, some researchers used electric fields as perturbations to adsorbate–surface model Hamiltonian without considering the Fermi energy of the electrode surface. This approach has become a school of thought different from or competing with Fermi energy shift perspective for treating electrode potential as done in the Gurney-Butler-Volmer approach. As discussed later, both schools of thought are only approximate and subject to criticism and the problems were overcome decades later using powerful computers and more complete and accurate Hamiltonians.

## 1.10 Parsons and Bockris Referenced Butler Model to the Potential of Zero Charge and Parameterized It with Chemical Bond Data

In 1951 Parsons and Bockris parameterized Born–Oppenheimer potentials for hydronium ion discharge over mercury at potentials measured relative to the potential of electrocapillary maximum, which is approximately the pzc [28]. What these authors did can best be understood as a modification of Fig. 1.4 where the electron energy has a new reference state corresponding to the pzc and the zero on the energy scale is redefined. From available bond data, they estimated the $W(r)$, $R(r)$, and $A(r)$ curves for $H^+$–$OH_2$, H–$OH_2$, and M–H bonds, respectively where M was Hg and Ni. They identified the H to M internuclear distance to be the thickness of the Helmholtz double layer, defined as the distance between the atoms of the charged metal and adsorbed solvated ions and presumed to be in the range 1.7–2.0 Å. Figure 1.6 illustrates the Helmholtz double layer in idealized form for a negatively charged electrode surface. This double layer is equivalent to a capacitor with the electric field increasing linearly from the negative layer to the positive layer. The idea was used in the Horiuti and Polanyi model of the previous section and other explanations based on the field model are discussed in later sections. Predictions from theoretical models that allow for non-uniform electrolyte ion distributions are also discussed later. The theories generally support double-layer structure and field models that were developed by numerous workers to give a more detailed description than is in the Helmholtz model.

The goal of Parson and Bockris was estimating the activation energy at the pzc. They stabilized the energy of the $H_3O^+$ system shown in Fig. 1.4 by the workfunction of the electrode at the pzc. The energy of free H atom + $H_2O$······M was defined as zero on the energy scale. The activation energy, $\boldsymbol{E_n}$*, for the pzc was then defined using the Franck–Condon principle as the difference in energy between the $H_2O$···H···M and the $H_2O$···$H^+$ + $e^-$ system in its ground vibration state. The relationship between this view and Figs. 1.4 and 1.7.

From the model, predicted activation energy over Hg was around 12 kcal/mol compared to an experimental estimate placing it in the range of about 14–19 kcal/mol and the predicted symmetry factor, α, was about 0.48 compared to the experimental estimate placing it in the range 0.5–0.55. Over Ni the activation energy was predicted as 6 kcal/mol and the experimental range was 7–10 kcal/mol. The

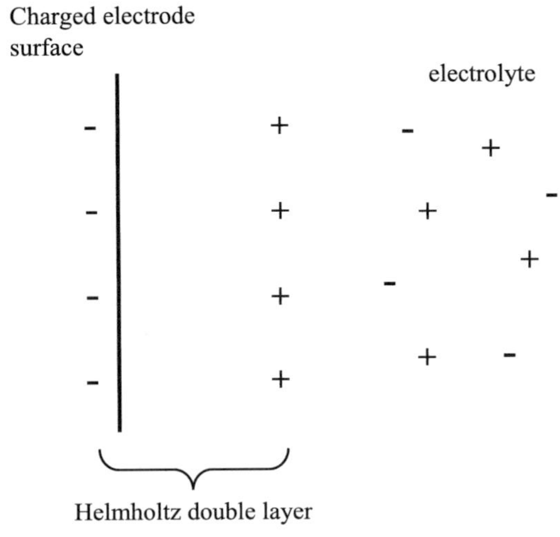


**Fig. 1.6** Schematic for the Helmholtz double layer in the case of negative surface charging and a layer of charge compensating anions a fixed distance from the surface. Cations and anions in the electrolyte further from the surface are randomly distributed. In this case the counter electrode (not shown) will have a balancing positive surface charge and a cation layer

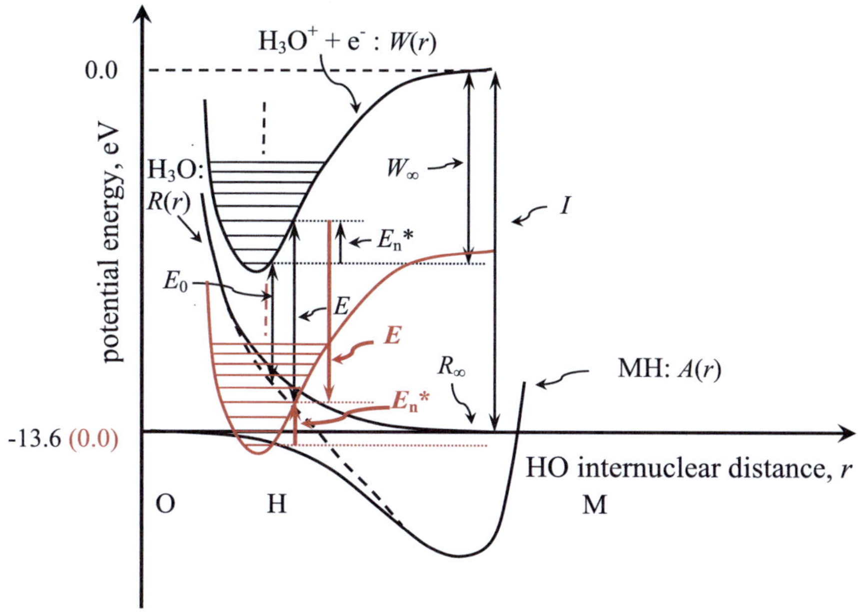


**Fig. 1.7** The relationship of the Parsons-Bockris (lower, in red) model to the Gurney-Butler theory (upper, in black) in Fig. 1.4. The energy *E* is the negative of the electrode workfunction at the potential of zero charge and is how much the Gurney-Butler *W*(r) curve is stabilized. $E_n$* is the electron transfer activation energy for the Parson-Bockris model and equals the energy of the crossing point of the O–H–M potential function with the shifted *W*(r) curve minus the ground vibrational energy

predicted symmetry factor was about 0.48 and the experimental range about 0.59–0.64. Although the theoretical model has approximations, the potential functions employed have approximations, and the experimental activation energies are approximate, the above comparison of the model based and experimental results suggests elements of generality and usefulness in the procedure. Parsons and Bockris also, using potential functions for $H_2O$ discharge, meaning reduction to form H(ads) + $OH^-$, predicted that the activation energy for $H_2O$ discharge would be 14 kcal/mole higher than the value predicted for $H_3O^+$ discharge to form H(ads).

## 1.11 Deficiencies of the Gurney-Butler and Related Models: Lack of Surface Charging and Double-Layer Polarization, and Others

This was the understanding of electron transfer and electrocatalysis in 1951. For numerous reasons, the Gurney-Butler treatment was approximate. Some on the approximations are listed.

1. The three-atom O–H–M potential is not really the sum of two pairwise Born–Oppenheimer potentials and the energy also depends on the O–H–M angle.
2. Modeling solvation stabilizations using a single hydrating water molecule is only partial.
3. When based on Born–Oppenheimer potentials, the electron affinity of the model does not lead to the correct electrode potential because it is an internal energy and not the Gibbs free energy.
4. For $H_3O^+$ transitions involving the tunneling tail of the proton vibrational wave-function into the $W(r)$ potential and the tunneling tail of the $H_2O$–H–M hydrogen wave function into the $A(d_{MO}-r)$ potential are neglected.
5. The assumed coupling of electronic states to nuclear position that gives rise to the Franck–Condon approximation during electron transfer, itself a tunneling process, will impart some error.
6. An accurate electron tunneling factor is difficult to determine.
7. Integration over the distance variable has not been accomplished due to lack of knowledge of details of electron and proton tunneling.
8. Changing the electrode potential by charging the electrode surface will affect structures and bond strengths between it and molecules adsorbed on it.
9. The double layer ion structure will change with electrode potential and will influence adsorbate structures and adsorption bond strengths.

Proton tunneling had already been considered using model potentials and it seemed possible that in cases not involving electron transfer there could be rate enhancement at laboratory temperatures, though parameterization of the model potentials was uncertain [29]. Despite all the approximations, the model is sufficiently robust that this qualitative picture remains a powerful conceptual tool and is an important basis for understanding.

## 1.12 More on the Importance of the Butler-Volmer Equation

The inability to calculate with adequate accuracy the Born–Oppenheimer potentials, electron affinities and ionization potentials associated with electron transfers hindered theoretical development from this point on until the 1990’s. Nevertheless, the well-known parameterized Butler-Volmer equation, which sums the currents for the forward and reverse reactions at a given potential in Eq. (1.9):

$$i = i_0[\exp(\alpha_a F\eta/RT) - \exp(-\alpha_c F\eta/RT)] \quad (1.9)$$

provides the operational framework for interpreting current measurements as functions of electrode potential and remains the basis for interpretation today. In this equation $i_0$ is the exchange current density, $\alpha_a$ and $\alpha_c$ are anodic and cathodic transfer coefficients respectively and $\eta$ is the activation overpotential, $U$–$U^0$. Transfer coefficients apply to many-electron reactions while the term "symmetry factor" is used for one electron reactions as in Figs. 1.5 and 1.6. As will be seen later, the computer programs for quantum mechanical calculations, developed later in the twentieth century, using realistic model systems for electron transfer and statistical theories, allow useful approximations of some of the parameters. Even more significantly, improved predictions extending beyond the Butler-Volmer framework become possible. As this began, Yeager stated in his excellent review in 1980 that "Electrochemistry is undergoing a renaissance, a key feature of which is the attracting of scientists to this area from other disciplines" [30]. But prior to that, additional concepts and models were advanced and they are presented next.

## References

1. Volta A (1800) On the electricity excited by the mere contact of conducting substances of different kinds. Phil Trans Royal Soc London 90:403–431
2. Nicholson W (1800) Account of the new electrical or galvanic apparatus of sig. Alex. Volta, and experiments performed with the same. Nicholson's J, J Nat Philos, Chem, Arts 4:179–187
3. For a review written two years before his death outlining many of Davys achievements (1826) In: Davy H, The Bakerian lecture. On the relations of electrical and chemical charges, Philosophical Transactions of the Royal Society of London vol 116, pp 383–422
4. Faraday M (1834) On electrical decomposition. In: Philosophical transactions of the royal society of London, Vol 124 pp 77–122
5. Jones B (1870) The life and letters of faraday, Vol I, Lippincott and Co., Philadelphia, p 310
6. See Nobelprize.org and *Nobel Lectures, Physics 1901–1921* (1967) Elsevier Publishing Company, Amsterdam
7. See Philosophical Magazine (1894) Series 5, Vol 38, p 418–420
8. Nernst W (1889) Die Elekrtomotorische Wirksamkeit der Jonen. Z Physik Chem 4:129–181
9. Walther N (2003) Reasoning of theoretical chemistry: nine papers (1889–1921) (Ger., Begründung der Theoretischen Chemie : Neun Abhandlungen, 1889–1921). Frankfurt am Main: Verlag Harri Deutsch
10. The Scientific Papers of J. Willard Gibbs Vol I (1906) Thermodynamics, Longmans, Green, and Co., London New York and Bombay
11. Gibbs J (1902) Willard elementary principles in statistical mechanics—developed with special reference to the rational foundation of thermodynamics. Charles Scribner's Sons, New York
12. Tafel J (1905) On the polarization during cathodic hydrogen evolution. Zeitschrift für physikalische Chemie, Stöchiometrie und Verwandtschaftslehre 50:641–712
13. Butler JAV (1924) Studies in heterogeneous equilibria. Part II. The Kinetic interpretation of the Nernst theory of electromotive force. Trans Faraday Soc 19:729–733
14. Butler JAV (1924) Studies in heterogeneous equilibria. Part III. A kinetic theory of reversible oxidation potentials at inert electrodes. Trans Faraday Soc 19:734–739
15. Butler JAV (1932) The mechanism of overvoltage and its relation to the combination of hydrogen atoms at metal electrodes. Trans Faraday Soc 28:379–382
16. Erdey-Grúz T, Volmer M (1930) Zur Theorie der Wasserstoffüberspannung. Zeit Phys Chem A 150:203–213
17. Erdey-Grúz T, Volmer M (1932) Zur Frage der Elektrolytischen Metallüberspannung. Zeit Phys Chem A 157(165):1931, 162, 53
18. Schrödinger E (1926) An undulatory theory of the mechanics of atoms and molecules. Phys Rev 28:1049–1070
19. Gurney RW (1931) The quantum mechanics of electrolysis. Proc Roy Soc 134A:137–154

20. Sommerfeld A (1928) Zur Elektronentheorie der Metalle auf Grund der Fermischen Statistik. 1. Allgemeines, Strömungs- und Austrittsvorgänge. Zeitschrift für Physik, Bd. 47:S 1–32
21. Sommerfeld A (1928) Zur Elektronentheorie der Metalle auf Grund der Fermi'schen Statistik. 2. Thermoelektrische, galvano-magnetische und thermomagnetische Vorgänge. Zeitschrift für Physik, Bd 47:S 43–60
22. Dirac PAM (1926) On the theory of quantum mechanics. Proc Roy Soc London, Ser A 112:661–677
23. Born M, Oppenheimer R (1927) Zur Quantentheorie der Molekeln. Annal Phys 84:457–484
24. Zhan C-G, Dixon DA (2003) The nature and absolute Hydration free energy of the solvated electron in water. J Chem B 107:4403–4417
25. Franck J (1926) Elementary processes in photochemical reactions. Trans Faraday Soc 21:536–542. Condon E. A theory of intensity distribution in band systems. Phys Rev 28:1182–1201
26. Butler JAV (1936) Hydrogen overvoltage and the reversible hydrogen electrode. Proc Roy Soc 157:423–433
27. Horiuti J, Polanyi M, Acta. Physiochimica U. R. S. S. 2(4) (1935) 505–532; English translation: "On the Theory of Proton Transfer. J Mol Catal A 199(2003):185–197
28. Parsons R, Bockris JO'M (1951) Calculation of the energy of activation of discharge of hydrogen ions at metal electrodes. Trans Faraday Soc 47:914–928
29. Bell RP (1933) The application of quantum mechanics to chemical kinetics. Proc Roy Soc London, Ser A 139:466–474
30. Yeager E (1980) Non-traditional approaches to the study of solid-electrolyte interfaces: problem overview. Surf Sci 101:1–22

# Chapter 2
# Born Model for Ion Solvation Energies

**Abstract** Born used a statistical model where the energy of interaction between ions and the surrounding solvent was approximated by the energy needed to charge ions modeled by spherical charge distributions in a dielectric solvent with its known dielectric permittivity. He assigned crystallographic radii to the ions and found the predicted solvation energies from groups ion the periodic table were made more accurate by adding 0.10 Å to the anion radii and 0.85 Å to the cation radii. This introduced semiempirical parameterization in electrochemistry.

## 2.1 General Introduction

Using quantum mechanics to parameterize the analytical formulas of Nernst and Butler-Volmer to include effects of the surrounding solution on the ions participating in electrochemical reactions was not possible until the development of powerful computers. However, models predicting solvation energies using classical physical chemistry were developed in the 1920s. The Born solvation model was successful for atomic ions and is reviewed in this chapter.

## 2.2 Born Theory for Ion Solvation Energy

In 1920 Born proposed a classical electrostatic ion solvation model for estimating the solvation stabilization that ions experience when immersed in dielectric solvent [1]. In the model an ion i of charge $q_\mathrm{i}$ is assumed to have a known radius $r_\mathrm{i}$ and is modeled as a conducting sphere with this radius. Coulomb's law gives the electrical work required to charge the sphere in vacuum when initially uncharged. For a sphere of charge $q$ surrounded by vacuum, the electric potential at the surface is $q(4\pi\varepsilon_0 r_\mathrm{i})^{-1}$, where $\varepsilon_0$ is the permittivity of vacuum, so the work required to add an increment d$q$ of charge to the charged sphere is $q\mathrm{d}q(4\pi\varepsilon_0 r_\mathrm{i})^{-1}$. Integrating over charge from zero charge to the ion charge $q_\mathrm{i}$ yields the charging work and Gibbs free

 
A. Anderson, *Past and Current Models in Electrochemistry*,
https://doi.org/10.1007/978-3-031-97791-6_2

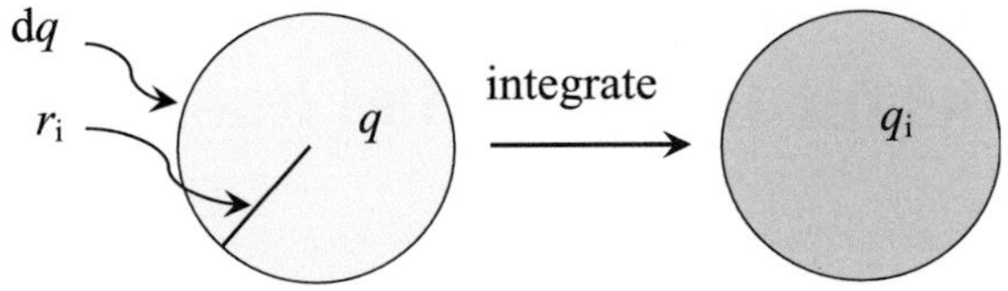


**Fig. 2.1** This illustrates the result of adding charge increments d$q$ to a conducting sphere of radius $r$ and charge $q$ on the left and integration results in accumulated charge $q_i$ shown on the right

energy value in Eq. (2.1):

$$G = (q_i e)^2 (8\pi \varepsilon_0 r_i)^{-1} \tag{2.1}$$

The process is illustrated in Fig. 2.1.

Textbooks on electricity and magnetism give the details for the following summary [2]. When immersed in a dielectric medium such as water the field of an object bearing charge $q$ induces an electric polarization dipole **P** in the medium. Gauss' law then takes the form of a surface integral:

$$\iint (\varepsilon_0 \mathbf{E} + \mathbf{P}) \cdot \mathbf{n} da = q \tag{2.2}$$

When the electric displacement is defined:

$$\mathbf{D} = \varepsilon_0 \mathbf{E} + \mathbf{P} \tag{2.3}$$

and the magnitude of the induced dipole is assumed linear in electric field amplitude:

$$\mathbf{P} = \chi \mathbf{E} \tag{2.4}$$

where $\chi$ is the electric susceptibility. Then when the dielectric's permittivity $\varepsilon$ is defined:

$$\varepsilon = \varepsilon_0 + \chi \tag{2.5}$$

and the relative permittivity or dielectric constant $K$ is defined:

$$K = \varepsilon / \varepsilon_0 \tag{2.6}$$

the electric displacement and electric field are

$$\mathbf{D} = q\left(4\pi r^3\right)\mathbf{r} \tag{2.7}$$

and

$$\mathbf{E} = q\left(4\pi K\varepsilon_0 r^3\right)^{-1}\mathbf{r} \tag{2.8}$$

The electrostatic energy for the charged sphere immersed in the dielectric may be evaluated by the equation

$$G = 1/2\int \mathbf{D}\cdot\mathbf{E}\mathrm{d}v \tag{2.9}$$

$$= (q_{\mathrm{i}}\mathrm{e})^2(8\pi\varepsilon r_{\mathrm{i}})^{-1} \tag{2.10}$$

where $v$ is the volume element and the integration is over the region external to the sphere. The permittivity of vacuum is 1.0 and for water the dielectric constant is about 80 at 20 °C and significantly temperature dependent, dropping, for example, to about 56 at 100 °C and for ice the value is about 92, increasing to about 106 at –50 °C [3]. As the temperature goes up, increased thermal motion lowers the polarization of the dielectric causing the decrease in the dielectric constant of water. The free energy change when the model ion in vacuum is transferred into the dielectric solution, i.e., the solvation energy, is then the energy for charging it in the dielectric minus the energy for charging the sphere in vacuum, Eqs. (2.1)–(2.10). The result is the Born equation:

$$\Delta G^0_{\mathrm{solv}} = (q_{\mathrm{i}}\mathrm{e})^2(8\pi\varepsilon r_{\mathrm{i}})^{-1} - (q_{\mathrm{i}}\mathrm{e})^2(8\pi\varepsilon_0 r_{\mathrm{i}})^{-1} \tag{2.11}$$

For a solute with large permittivity like water the first term in this equation makes scant contribution to the solvation energy.

Equation (2.11) was derived assuming the ion is a conducting sphere, that it is not polarizable, and that molecular details of the dielectric can be approximated by a continuum model, meaning the molecules, such as water, are point dipoles but do not participate in bonds to one ion that are different from bonds to another ion and, essentially, the interfacial chemical bond energy is zero. That is, all the energy of solvation comes from the bound charge of the aligned dipoles in the dielectric continuum. This bound charge essentially neutralizes part of the free charge at the surface of the spherical conducting model for the ion. The model does not include the increase in free energy that accompanies the change from random dipole orientation to ordered dipoles when the conductor is charged. In fact, this classical model, dependent on rigid charges and continuum dielectric constants, by definition does not lend itself to improvements via quantum chemical modeling with explicit water molecules. However, as shown in a later chapter, a hybrid approaches using explicit molecules with and without the dielectric constant of bulk water can make good predictions of solvation energies. Despite the simplifications in the classical approach, when using crystallographic radii resulting trends in solvation energies follow experimental measurements. This theory allows introduction of semiempirical parameters for improved agreement of predicted solvation energies with measured

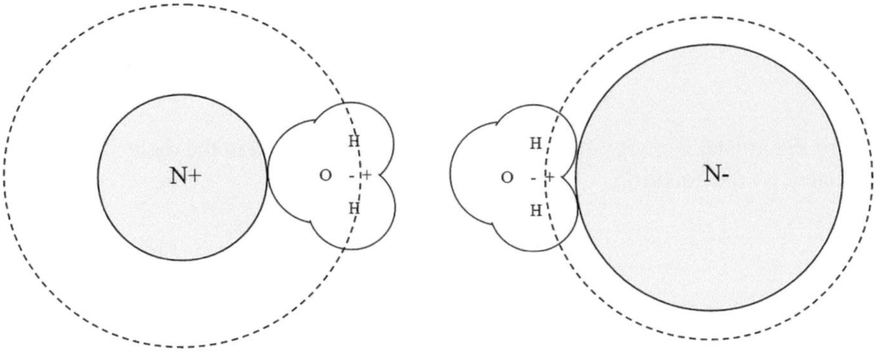


**Fig. 2.2** The figure on the left shows a water molecule in the inner solvation shell coordinated to a cation through the O atom giving the cation a new larger radius, specified by a dashed line circle, that extends to the center of the water electric dipole. On the right the water molecule is coordinated to an anion through the H atoms and a dashed circle passes through the middle of the water dipoles. The dashed circles represent the effective ion radii used by Latimer et al. in the Born equation

values. Latimer et al. [4] found that adding 0.10 Å to the crystallographic radii of $F^-$, $Cl^-$, $Br^-$, and $I^-$ to the corresponding crystallographic radii as defined by Pauling[5] led to agreement with the measured solvation energies from the Born equation. In the case of the alkali metal cations, adding 0.85 Å to all the crystallographic radii accomplished the same improved fit to experiment. This was rationalized by defining the radius of the charged sphere to equal the sum of the crystallographic radius and the distance to the center of the water dipole, as indicated in Fig. 2.2. This is done so that

the surface used in the development of Eq. 2.8 includes both the free charge at the conductor surface, $N^+$ or $N^-$, and the bound charge at the interface, which is charge of dipole ends contacting the ion surface.

Success of the semiempirical fitting, by adding 0.10 Å to the anion radii and 0.85 Å to the cation radii, these radii being based on crystal structure data, has inspired refinements up to this day. It is notable that the model, which uses only classical electrostatics and no bonding information, either empirical or from quantum mechanical calculations, is so successful.

## References

1. Born M (1920) Volumen und Hydratationswärme der Ionen. Z Phys A: Hadrons Nucl 1:45–48
2. Reitz JR, Milford FJ (1960) Foundations of electromagnetic theory, Addison-Wesley, Reading MA
3. Lide DR (ed) (2005) Handbook of chemistry and physics, 86th edition. CRC Press, Boca Raton FL
4. Latimer WM, Pitzer KS, Slansky CM (1939) The free energy of hydration of gaseous ions and the absolute potential of the normal calomel electrode. J Chem Phys 7:108–111
5. Pauling L (1927) The sizes of ions and the structure of ionic crystals. J Am Chem Soc 49:765–790

# Chapter 3
# Models of the 1950s and 60s

**Abstract** Following the Born model for ion solvation energies in 1920 there was a laps in model development until Libby attempted to explain electron exchange reactions by invoking Born–Oppenheimer bond vibrational potentials and the Franck–Condon factor for electron transfer in a way that did not conserve energy. It is shown how Libby might have succeeded had he invoked electrochemistry by using a half-cell reaction path for the electron exchange. Marcus noted Libby's failure to conserve energy and developed a model for energies of electron transfer transition states where there are two kinds of polarizations of the dielectric, one dependent on the electric field the ion experiences from the other ion and a structural one that does not. Marcus introduced the ion solvation shell and used the breathing mode model of George and Griffiths satisfy the Franck–Condon factor. The Marcus theory identified "inverted regions" where a reaction would be activated though the product is more stable than the reactant. Hush developed a theory for electron transfer at electrode surfaces where at the transition state the electron is delocalized over the reactant and transition state. Both the Marcus and Hush theories gave electron transfer parameters for use in the Butler-Volmer equations.

## 3.1 General Introduction

The big development during this period was the Marcus theory for electron transfer between ions in solution and at interfaces. It depended on a statistical model for ion interaction with the solvent with its dielectric constant. For his complete theory, which is used widely today, he added Born–Oppenheimer vibrational potential energy functions for the chemical bonds and this led to the famous and widely used Marcus diagrams. In the same time period Hush proposed a model where an electron transferring adiabatically is delocalized over reactant, transition state, and product structures. He showed that within the model electron transfer activation energies could be expressed as functions of solvation energies of reactant, product, and transition state.

 
A. Anderson, *Past and Current Models in Electrochemistry*,
https://doi.org/10.1007/978-3-031-97791-6_3

## 3.2 Libby's Model for Electron Exchange Reactions

In 1952 Libby proposed a model intended to explain activation energies for electron exchange reactions in aqueous solution [1]. In his description, the ions, which included a hydration atmosphere made up of water molecules, with structures not known in detail, would change following electron transfer. An electron exchange reaction between ions at equilibrium has the form

$$M_l^{n+} + M_r^{(n-1)+} \rightleftharpoons M_l^{(n-1)+} + M_r^{n+} \tag{3.1}$$

where ion positions are distinguished by the subscripts l for left and r for right to aid the following discussion.

Libby envisioned a Franck–Condon process for the electron transfer as jumping from the Born–Oppenheimer potential energy function of the $M^{(n-1)+}$ ion with its stable hydration structure to the Born–Oppenheimer potential energy function of the $M^{n+}$ in its hydration structure. He noted that: "Of course there is no net energy change in the reaction" [1]. Let us analyze this by supposing that, of all structure coordinates, the distance to the first water shell most strongly affects the hydration energy of each ion. As shown in Fig. 3.1, at equilibrium that distance will be shorter for the more positively charged $M^{n+}$ ion.

Let R be the distance from the ion center to the first hydration shell. The potential curve will be deeper and narrower for the more highly charged ion, $M^{n+}$ because of stronger bonding between the water molecules and the ion. Now suppose the electron

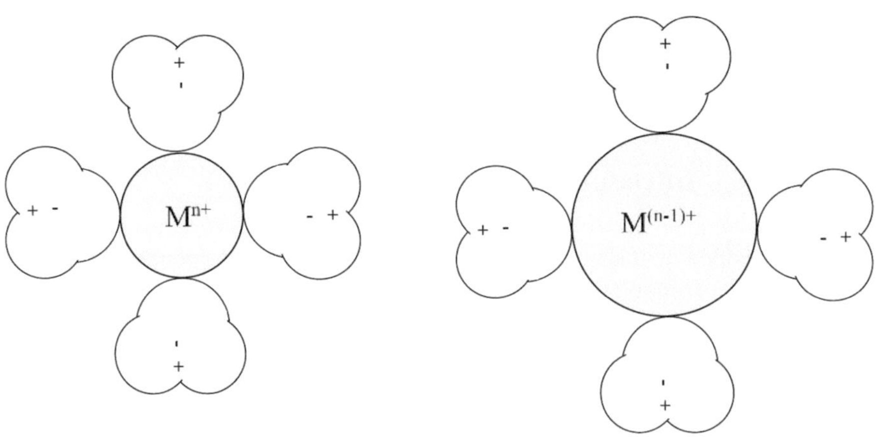


**Fig. 3.1** Schematic cross section view of the expansion of first hydration shell after the charge on an ion decreases from n+ to (n–1)+ . On the left a cation with charge n +is represented by a circle with four water molecules coordinated to it in the plane through the O atoms. On the right the cation charge is reduced to (n–1)+ and is draw with a larger radius with the four water molecules coordinated at the new radius. Each water dipole included

is removed from the $M_l^{(n-1)+}$ ion to transform it into $M_l^{n+}$. Applying the Franck–Condon principle for a sudden electron transfer, meaning with fixed structure, as in Fig. 3.2, shows $M_l^{n+}$ is not relaxed to its equilibrium structure. The $M^{n+}$ ion forms in an excited vibrational state, as shown in Fig. 3.3. This will not be a spontaneous physical occurrence without energy input, say in the form of light. To overcome the need to input energy, Libby asserted that observed rapid electron exchange reactions "can be catalyzed by complexing the exchanging ions in such a way that the complexes are symmetrical providing their geometries are identical to within the vibration amplitudes involved in zero-point motion." Such a symmetrical arrangement would require distortions of both complexes requiring energy input but would contradict the assumption that the exchange reaction occurs with no net energy input.

If the $M_l^{n+}$ ion gains an electron suddenly, Fig. 3.1 then takes the form of Fig. 3.3. This time the ion that is formed suffers compression by the unrelaxed hydration shell. The result is the $M_l^{(n-1)+}$ is formed in an excited vibrational state as shown in Fig. 3.5.

Libby's model did not take into account that excited molecular vibrational states are formed following transfer of the electron. Libby assumed there would be little or no barrier for Frank–Condon electron transfer, and attributed the activation barriers in electron exchange reactions to repulsions between the approaching positively charged solvated ions. He hypothesized the repulsion could be reduced by intervening anions which activated the electron transfer and he applied the classical Born model to calculate interaction energies of the positively charged ions with the solvent. The effective dielectric constant was noted to be ill-defined at the molecular scale

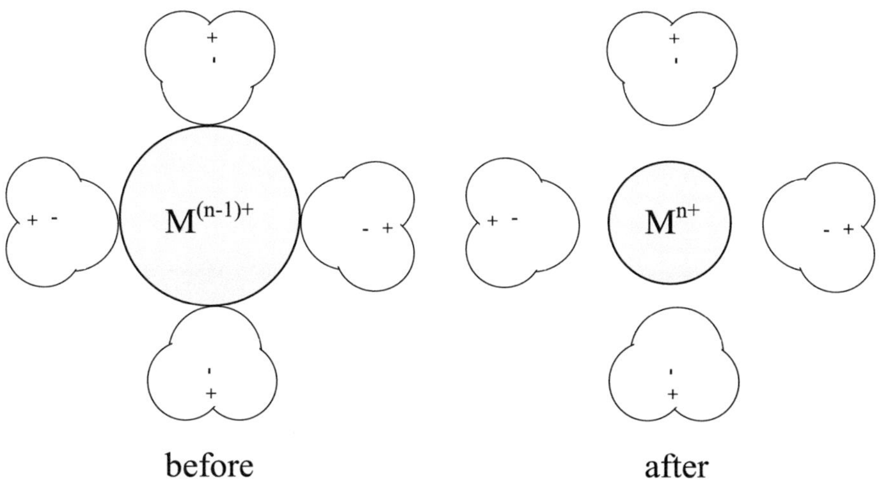


**Fig. 3.2** On the left to the diagram is the structure of $M^{(n-1)+}$ before the electron transfer and it is the same structure as in the right hand picture in Fig. 3.1. On the right is a picture after electron transfer forming $M^{n+}$ according to the Franck–Condon principle where the radius of the ion is reduced while the water molecules are unchanged in their positions

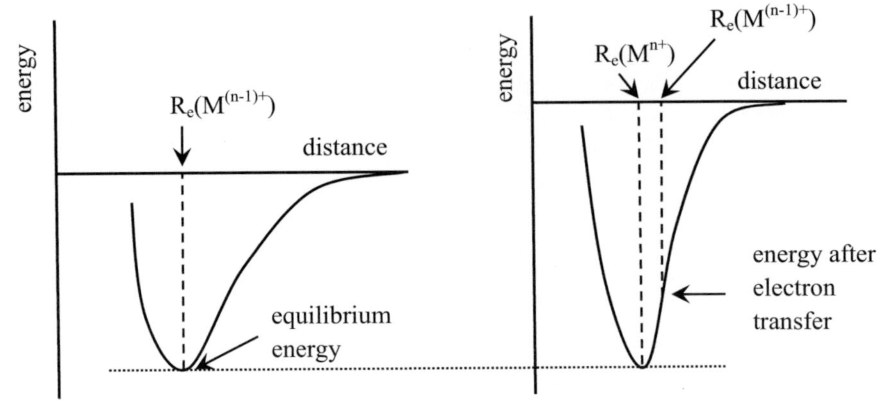


**Fig. 3.3** The Morse-shaped potential curve on the left is for the hydration shell at a distance $R(M^{(n-1)+})$ from the $M^{(n-1)+}$ ion with equilibrium distance $R_e(M^{(n-1)+})$ from the ion center to the O atoms and, similarly, the curve on the right is for the $M^{n+}$ ion with equilibrium distance $R_e(M^{n+})$. The energy at $R_e(M^{(n-1)+})$ on the $R(M^{n+})$ curve corresponds to a vibrational excited state for the n+ ion

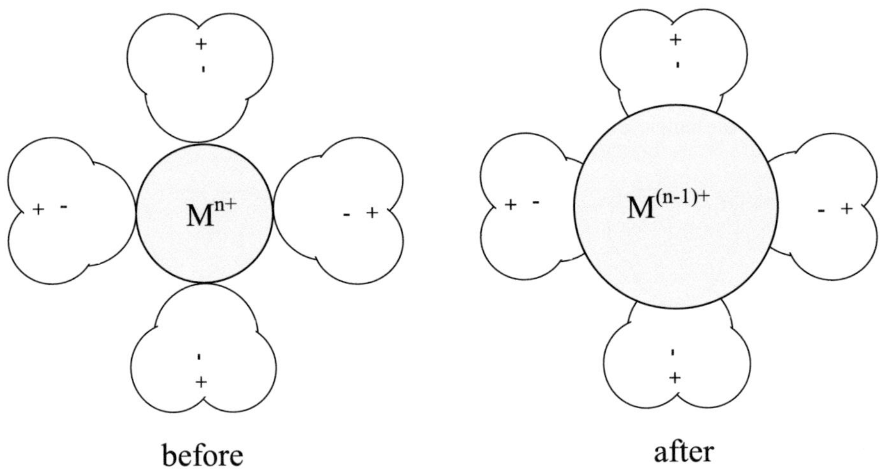


**Fig. 3.4** On the left is the structure of $M^{n+}$ before the electron transfer and it is the same structure as in the left hand picture in Fig. 3.1. On the right is a picture after electron transfer forming $M^{(n-1)+}$ according to the Franck–Condon principle where the radius of the ion is increased while the water molecules are unchanged in their positions

and the energetics of counter anion-cation interactions were not addressed, though Debye Hückel and had provided a formula in 1923 for the energies of individual ions in solution [2, 3]. No quantitative theory came from Libby's suggestions, but the Franck–Condon principle continued to play a role in discussions of electron transfers between solvated ions.

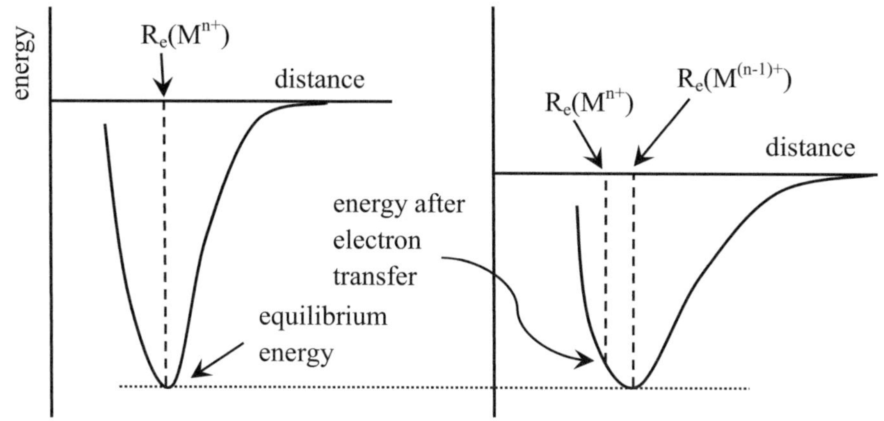


**Fig. 3.5** The Morse-shaped potential curve on the left is for the hydration shell at a distance $R(M^{n+})$ from the $M^{n+}$ ion with equilibrium distance $R_e(M^{n+})$ from the ion center to the O atoms and, similarly, the curve on the right is for the $R(M^{(n-1)+})$ ion with equilibrium distance $R_e(M^{(n-1)+})$. The energy at $R_e(M^{n+})$ on the $R(M^{(n-1)+})$ curve corresponds to a vibrational excited state for the (n−1)+ ion

The Libby model consists in switching from one potential curve to another without consideration of balancing charge and mass. It was shown in Fig. 1.5a in Chap. 1 how Butler's potential models, which include the energies of electrons that balance charge and mass, provide an understanding of potential dependencies of electron transfer activation energies when using rigid Born–Oppenheimer molecular potential energies and a fixed O-M distance. It was seen in Fig. 1.5b that harder collisions of $H_3O^+$ with M at the reversible potential will lower the electron transfer activation energy to zero but at H–O distances less than the M–O equilibrium value the activation energy increases.

## 3.3 Beyond Libby: Glimpsing into the Future by Treating the Reduction of $M^{(n-1)+}$ and the Oxidation of $M^{n+}$ as Electrochemical Half-Cell Reactions, Where M Stands for Molecule

An alternative viewpoint, presented in detail later, is to calculate activation energies for the oxidation and reduction reactions as functions of electrode potential. Let the species to be reduced, in this case $(H_2O)_4M^{n+}$ be in its lowest energy structure, called the reduction precursor structure. This has an electron affinity $EA_{rp}$ and when the electrode is set to a potential matching the electron affinity, $U_{rp}$, reduction can take place with zero activation energy. One can write:

$$(H_2O)_4M^{n+}_{(red)} + e^-(U_{rp}) \rightleftharpoons (H_2O)_4M^{(n-1)+}_{(red)} \quad (3.2a)$$

Here the subscript (red) indicates the optimized reduction precursor structure. The activation energy increases from zero when the distance between the water molecules and $M^{n+}$ increases or decreases. Each distance defines a structure with a new reduction potential. The reduced molecule on the right hand side of Eq. (3.2a) is not in its most stable structure and since M bears one less positive charge the water molecules will move away forming a new equilibrium structure called the oxidation precursor with an ionization potential $IP_{op}$ and its oxidation potential is $U_{op}$. For this one writes:

$$(H_2O)_4M^{(n-1)+}_{(ox)} \rightleftharpoons (H_2O)_4M^{n+}_{(ox)} + e^-(U_{op}) \quad (3.2b)$$

The subscript (ox) indicates the optimized oxidation precursor structure. The activation energy is zero for this equilibrium and increases from zero as the distance between the water molecules and $M^{(n-1)+}$ increases or decreases and each distance defines a structure with a new oxidation potential.

The reduction and oxidation activation energies can be plotted as functions of the applied electrode potential. Where they cross marks the reversible potential and the activation energy for the equilibrium electron exchange reaction. This is depicted in Fig. 3.6.

It is shown later that the reversible potential equals the $(E_{rp}-E_{op}-\varphi)$V/eV where $E_{rp}$ and $E_{op}$ are the energies calculated for the reduction and oxidation precursor systems and $\varphi$ is the thermodynamic workfunction of the standard hydrogen electrode, when all are reported in eV units. It is also shown that at any given potential $U$, $E_a(\text{ox},U) = E_a(\text{red},U) + E_{rp}-E_{op}-(\varphi + U/\text{V})$ eV. This means everything is derivable when either the reduction activation energy curve or the oxidation activation energy curve are known.

Applying the above formalism depends on the ability to calculate the energy and electron affinity of the reactant as a function of its structure, and in this case the structure parameter is simply the distance between the metal ion and the solvating shell of water molecules. At the time of Libby's work computational theory had not advanced to the stage where energies of oxidation and reduction precursors or the electrode potential-dependent activation energies could be easily calculated. Lacking the computational ability, workers at the time did not consider the possibilities such as outlined above that become feasible decades later. So instead the focus of electrochemists continued to be classical models that did not anticipate the benefits of modern quantum chemical computations.

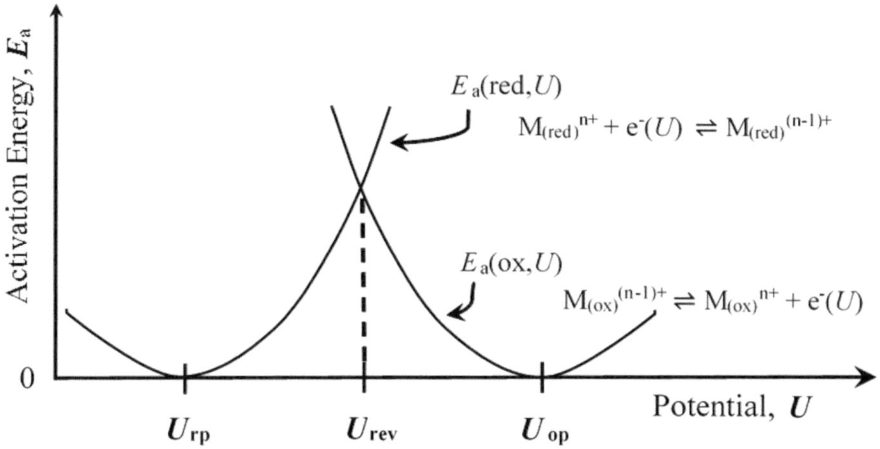


**Fig. 3.6** The energy versus potential parabola on the left is for Reaction 3.2(a) and has a minimum at the reduction precursor potential, $U_{rp}$, which is the electron potential when an electron transfers spontaneously to the species to be reduced, which is in its equilibrium structure. At higher and lower potentials, the parabola represents the activation energies for changing the structure of the precursor so that its negative electron affinity equals the selected electron potential. Potential increases going to the right. The parabola on the right is for Reaction 3.2(b) and has its minimum at the oxidation precursor potential, $U_{op}$, and this parabola represents the activation energies for reaching structures where oxidation at a chosen potential can occur. The curves cross at the equilibrium potential where oxidized and reduced species are in equilibrium

## 3.4 Marcus Approach to Electron Exchange Reactions

In 1956 R. A. Marcus saw that the Libby proposal did not conserve energy, and he set out to develop an electron transfer theory using hydrated ions and electrostatic interactions with the surrounding medium in which energy would be conserved [4]. He introduced a model for electron exchange reactions in aqueous solution wherein the first hydration shells were taken as rigid and both $M^{n+}$ and $M^{(n-1)+}$ were taken to have the same radius, a, which was the sum of an assigned ionic radius with the hydration shell included [5]. The model, developed in a series of papers over a period of nine years, formed like a slowly developing photographic print. In the first paper he derived formulas for the case of electron exchange between elementally different ions with radii $a_1$ and $a_2$. The Gibbs energy for the reacting system then depended on the interactions of the charged spheres with molecules outside the spheres and the Coulomb repulsion between the spheres, $q_1q_2e^2/(4\pi er)$ where $r$ is the distance between the centers. The charge distribution for the case of zero interaction was used, namely the charge distribution in either the reactant state or the product state. The surrounding dielectric medium was then assumed to interact with, for example, the $[M_l^{n+}, M_r^{(n-1)+}]$ system at the transition state to produce the remaining energy through two components of polarization:

$$\mathbf{P}(\mathbf{r}) = \mathbf{P}_e(\mathbf{r}) + \mathbf{P}_u(\mathbf{r}) \tag{3.3}$$

The first, or $E$-type, is an electronic polarization that varies with the electric field strength at position $\mathbf{r}$ and the second, or $U$-type, is an atomic plus orientation polarization that, during a rapid change in electronic charge distribution, as accompanies an electron transfer reaction, is independent of the electric field strength. The $E$-type polarization depends on the flexible component of the electron distribution function while the $U$-type polarization does not. At the transition state structure for the electron transfer $E$-type polarization changes when the electron transfers but $U$-type does not because the structure before and after the electron transfer is unchanged. The $U$-type polarization included multipole fields from the hydration shell and monopole fields from the counter ions. The transition state was taken to be defined by the polarization for which, for a given value of inter-ion distance R, the energy of the reactant electronic state equaled the energy of the product electronic state. The Gibbs energy of the transition state for this model before electron transfer, $G^*$, was, as shown by Marcus to be given as [6]:

$$G^* = 1/2\int \left[\mathbf{E}_c^{*2}/4\pi - \mathbf{P}^*\mathbf{E}_c^* + \mathbf{P}_u\left(\mathbf{P}_u/\alpha_u - \mathbf{E}^*\right)\right]\mathrm{d}v \tag{3.4}$$

Here $\mathbf{E}_c{}^*$ is the field strength the charged spheres would create if in vacuum, $\alpha_u$ is the susceptibility associated with $U$-type polarization, and d$v$ is the volume element. After electron transfer, the formula for the Gibbs energy $G$ at the transition state is similar:

$$G = 1/2\int \left[\mathbf{E}_c^2/4\pi - \mathbf{P}\mathbf{E}_c + \mathbf{P}_u(\mathbf{P}_u/\alpha_u - \mathbf{E})\right]\mathrm{d}v \tag{3.5}$$

The relationship between $G^*$, $G$, and the standard free energy of reaction, $\Delta G^0$ is

$$\Delta G^0 = \left(G^* - W_{iso}^*\right) - T\Delta S_e - (G - W_{iso}) \tag{3.6}$$

Here $W_{iso}{}^*$ and $W_{iso}$ are the Born solvation energies of the isolated ions in solution and $\Delta S_e$ is the change in electron entropy due to changes in electronic degeneracy. Pictures of the process are helpful. Figure 3.7 shows the model and states of the ions and Fig. 3.8 shows the energy level diagram.

In Eq. (3.3) and (3.4) $\mathbf{P}_u$ is undetermined and Marcus used variation theory to minimize $G^*$ and $G$ to show that

$$\int \left(\mathbf{P}_v/\alpha_u - \mathbf{E}^*\right)\delta\mathbf{P}_u\mathrm{d}v = 0 \tag{3.7}$$

and

$$\delta G^* - \delta G = \int \left(\mathbf{E} - \mathbf{E}^*\right)\delta\mathbf{P}_u\mathrm{d}v = 0 \tag{3.8}$$

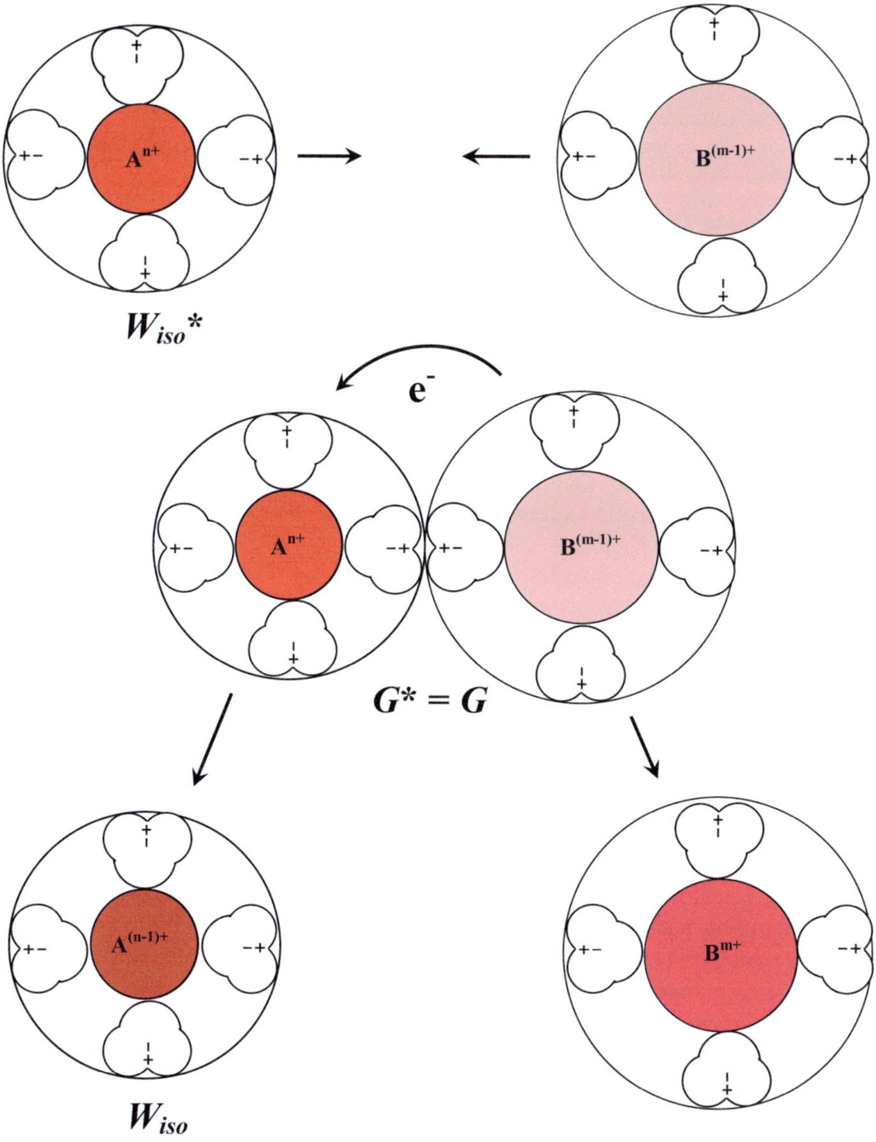


**Fig. 3.7** At the top, a schematic of reacting ions $A^{n+}$ and $B^{(m-1)+}$ with their solvation shells coming together; below in the middle the ions touch and an electron transfers from $B^{(m-1)+}$ to $A^{n+}$; below this at the bottom the then separation to $A^{(n+1)+}$ and $B^{m}$ product ions, bottom. The diameter of A with its hexaquo hydration is smaller than that of B with its hydration shell and they are fixed during the electron transfer in this model

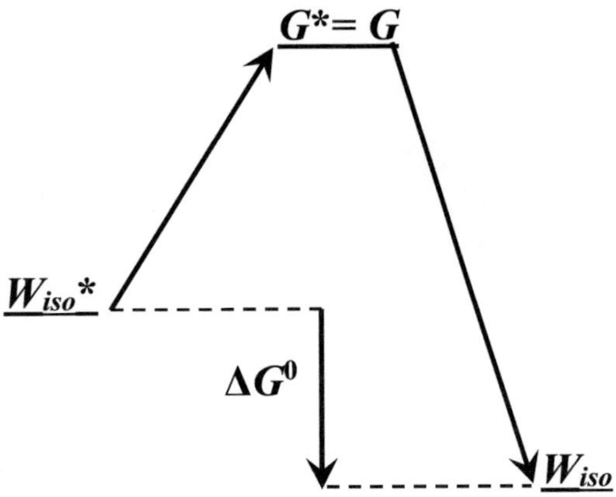


**Fig. 3.8** Graph of the electrostatic (Gibbs) energies of isolated reactant, energy $W_{iso}$*, transition state, energy $G^* = G$, and isolated product, energy $W_{iso}$*. The entropy term of Eq. (3.6) is not included but it will stabilize or destabilize $W_{iso}$ slightly

Applying the Lagrange method of undetermined multipliers led to the solution, which is most conveniently expressed as [7]:

$$\Delta G^* = m^2\lambda + e_1^* e_2^* e_0/e_s r \tag{3.9}$$

Here m is the Lagrangian multiplier and is in the expression

$$2m + 1 = -\left[\Delta G^0 + (e_1 * e_2 * - e_1 e_2)e_0/e_s r\right]/\lambda \tag{3.10}$$

where

$$\lambda = \left[1/(2a_1 + 1/2a_2 - 1/r)\left(e_0/e_{op} - e_0/e_s\right)(\Delta e)^2\right. \tag{3.11}$$

In these equations a1 and a2 are effective radii of the reactants, $e_s$ is the static component of the permittivity, $e_{op}$ is the optical component, and $\Delta$e is the number of electrons transferred between the ions. The term $m^2\lambda$ is a component of the activation energy of charge transfer. The other component $e_1{}^*e_2{}^*e_0/e_s r$ in Eq. (3.9) is an energy difference for bringing together the two reacting ions from a large separation and no interaction. The ion energies at large separation do not include interactions with the ionic atmosphere so that the $e_1{}^*e_2{}^*e_0/e_s r$ component is accurate only in the limit of low ionic strengths, a point Marcus noted later [8].

## 3.5 George and Griffith: Introducing of the Breathing Mode for First Ion Hydration Shell

In the meantime, George and Griffith proposed that for exchange reactions between ions in solutions the activation energy for electron transfer could be expressed, at least approximately, in terms of the breathing force constants for the hydration shells [9]. This was done as follows.

Born–Oppenheimer potentials contain spectroscopic constants used in analysis of vibrational spectra when the potentials themselves are written as a Taylor series about the equilibrium internuclear distance, $R_e$:

$$E(R - R_e) = \text{Const.} + k_e/2!(R - R_e)^2 + l_e/3!(R - R_e)^3 + m_e/4!(R - R_e)^4 + \cdots \quad (3.12)$$

The $k_e$, $l_e$, and $m_e$ coefficients are, respectively, the harmonic, cubic, and quartic force constants and, since the nuclear repulsion component of the potential energy is singular as $R \to 0$, the radius of convergence for the series is $R_e$. The $l_e$ value is usually negative and the cubic term is sometimes assumed to relate to the depth of the potential, that is, the bond dissociation energy. However, George and Griffith restricted their analysis to the harmonic level.

The model proceeds as follows, using hexaquo $M^{2+}$ and $M^{3+}$ for illustration. Let the breathing mode force constant for $M^{2+}$ be $k_{e2}$ and for $M^{3+}$ be $k_{e3}$ and let the equilibrium bond lengths be $R_{e3}$ for $M^{3+}$ and $R_{e2} = R_{e3} + a$ for $M^{2+}$. The two non-interacting hydrated ions are in proximity to allow electron transfer. Let the harmonic force constant for the six ligands vibrating in concert in the breathing mode be six times the $M^{n+}$–$OH_2$ stretching force constant. Then, the energy $E_A$ of the pair in the transition state structure with $R_{e3}$ displaced by α and $R_{e2}$ displaced by β before electron transfer, ignoring the easily added zero point energies, is

$$E_A = 3\left[k_{e3}(\alpha)^2 + k_{e2}(\beta)^2\right] \quad (3.13)$$

After electron transfer, the potential functions switch position and the energy $E_B$ is given by

$$E_B = 3\left[k_{e2}(\alpha)^2 + k_{e3}(\beta)^2\right] \quad (3.14)$$

Figure 3.9 shows corresponding Born–Oppenheimer potentials which are unsymmetrical about the equilibrium, but harmonic in the limit of small displacements. The sum of their displacements is.

$a = \alpha + \beta$ and the components of $E_A$ and $E_B$ are the sums of the energies at the intersections (represented by the dotted vertical lines with the potential curves) of distance α from $M^{3+}$ or distance β from $M^{2+}$. When the constraint $E_B = E_A$ is imposed to conserve total energy, as was done by George and Griffiths, it follows that, for a given value of a, the potentials must be symmetric with $\alpha = a/2$, $k_{e2} = k_{e3}$, and

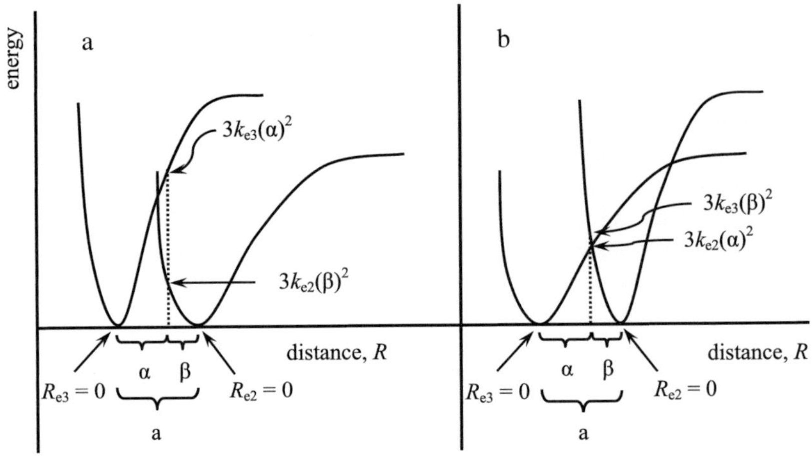


**Fig. 3.9** Part a is a graph of overlapping breathing shell Born–Oppenheimer potentials for $M^{3+}$ (left) and $M^{2+}$ (right) as functions of the O-M distances. Displacement of the shell about $M^{3+}$ from equilibrium is α and about $M^{2+}$ it is β. In the harmonic approximation, the energy of the $M^{3+}$ curve at α added to the energy of the $M^{2+}$ curve at β is $E_A$ in Eq. (3.13). The sum of α and β is a and the minima of the potentials are separated by a. Part b has the potential curves interchanged after electron transfer and crossing at $E_B$ which is given in Eq. (3.14)

$E_A = (3/2)k_{e2}a^2$. However, in general, $E_A \neq E_B$, and imposing the Franck–Condon principle then leads to the problems of the Libby model discussed earlier.

For the harmonic potentials the energy conservation problem is overcome by finding the structure where $E_A = E_B$. The displacement $\alpha_A$ for which $E_A$ is has a minimum value can be determined by the stationary point condition $\partial E_A/\partial\alpha_A = 0$:

$$\alpha_A = ak_{e2}/(k_{e2} + k_{e3}) \tag{3.15}$$

The minimum energy $E_{Amin}$ is gotten by substituting $\alpha_A$ in Eq. (3.12):

$$E_{A\ min} = 3a^2k_{e2}k_{e3}/(k_{e2} + k_{e3}) \tag{3.16}$$

An equation identical to Eq. (3.16) holds for $E_{Bmin}$ and because of this symmetry $E_{Amin}$ is a thermal activation energy and there is no need for electron de-excitation to conserve energy following the Franck–Condon state change.

## 3.6 Marcus Completes His Theory by Incorporating the George-Griffith Thermal Activation

Marcus incorporated a generalization of the George-Griffith model for thermal activation into his work and thereby brought the hydration shell structure and energy into the Gibbs activation energy. From the structures he introduced the concept of reaction coordinate [10]. He used the charge distribution in the zero interaction case, that either of the reactant or the product, in his derivation. The net result adds a term $\lambda_i$ to $\lambda$ in Eq. (3.11) so that the new $\lambda$ is

$$\lambda = \lambda_0 + \lambda_i \tag{3.17}$$

and $\lambda_0$ is the $\lambda$ in Eq. (3.11). Then when the multi-coordinate vibrational potential is expressed to the harmonic level on coordinate displacements $q_j$–$q_j{}^o$ with force constants $k_j{}^r$ and $k_j{}^p$ for reactant and product, respectively, then the parameter $\lambda_i$ is given by [11]:

$$\lambda_i = \sum_j \left(k_j^r k_j^p\right)\left(k_j^r + k_j^p\right)^{-1}\left(q_{jp}^o - q_{jr}^o\right)^2 \tag{3.18}$$

where $q_{jp}{}^o$ is equilibrium internal coordinate j for the product and $q_{jr}{}^o$ is that for the reactant.

For a single vibrational mode, this is the same form as Eq. (3.16) and follows the George-Griffith model.

Equation (3.9) is the statement of activation Gibbs energy and using the generalized $\lambda$ from Eq. (3.17), it takes the form

$$\Delta G^* = m^2(\lambda_0 + \lambda_i) + e_1 * e_2 * e_0/e_s r \tag{3.19}$$

For a given inter-ion distance, $r$, and fixed value for $m$, the activation Gibbs energy increases as $\lambda$ increases. The parameter $\lambda$ relates the activation energy to the reorganization of the solution as modeled by the dielectric continuum in the $\lambda_0$ term and to the reorganization of the hydration shells in the $\lambda_i$ term. The more the solvent reorganizes in the transition state, the larger the activation Gibbs energy and the more the hydration shells reorganize, the larger the activation Gibbs energy. However, $m$ depends on $\lambda$, (Eq. 3.10), and it would seem difficult to discover a simple picture for making predictions. In his part-four paper in the series, Marcus introduced approximations that provided a simpler and more easily applied model [10]. He imposed a linear response condition by assuming the dielectric polarization of the solvent was proportional to the change in charge on the reactants. He then expressed $G_r$ (for the reactant) and $G_p$ (for the product) as Taylor series about their minimum energies along a reaction coordinate and truncated the series at the quadratic level. He then showed that the resulting parabolas were rigid in shape and described by

$$G_r = (K/2)q^2 \tag{3.20}$$

for the reactant and by

$$G_p = \Delta G^0 + (K/2)(q - a)^2 \tag{3.21}$$

for the product, where q is the reaction coordinate and a is the separation between the minima of the parabolas. Let the value of q at the crossing point be α. Then

$$\alpha = \Delta G^0/(Ka) + a/2 \tag{3.22}$$

and

$$\Delta G_r^* = (K/2)a^2m^2 \tag{3.23}$$

where

$$m = -\Delta G^0/\left(Ka^2\right) - 1/2 \tag{3.24}$$

The model is applicable when the overlap of the product electronic wavefunction with the reactant electronic wavefunction is small so that there is a small quantum mechanical splitting at the crossing points of $G_r$ and $G_p$. This allows adiabatic electron transfer between the reactant electronic state to the product electronic state in accordance with the Franck–Condon principle, meaning the system has significant probability for entering the product state. This comprises the widely known Marcus theory for electron transfer, expressed with pictures in Fig. 3.10.

Each ion exchange reaction in solution would have its m value. The inverted region shown in Fig. 3.10 is a structure where $\Delta G_r$* is greater than zero even though the product is more stable than the reactant. The product curve could move up and down if the product of the reaction were an oxidized ion and the electron were given the energy $E_f$ of the electrode to which it transferred. For reduction the electron would originate at the electrode surface $E_f$. This is in accordance with the Gurney theory, though at this development stage Marcus seemed unaware of Gurney's and Butler's work because their studies were not cited. The experimental realization of the inverted region for ions came years later from measurements of intramolecular electron transfer rate constants for a series of molecules [12]. This proof of concept contributed to the Nobel Committee's decision to award Marcus a Nobel Prize.

Full details of the overall model were published in 1965 [13]. The application to the electrochemical interface, i.e., to oxidation–reduction reactions at electrode surfaces, introduced in parts V and VI, is straightforward. For electron transfers from or to a metallic electrode the density of filled and empty electrode states will eliminate the inverted region, as will be shown later. An excellent phenomenological presentation may be found in a frequently cited review [14].

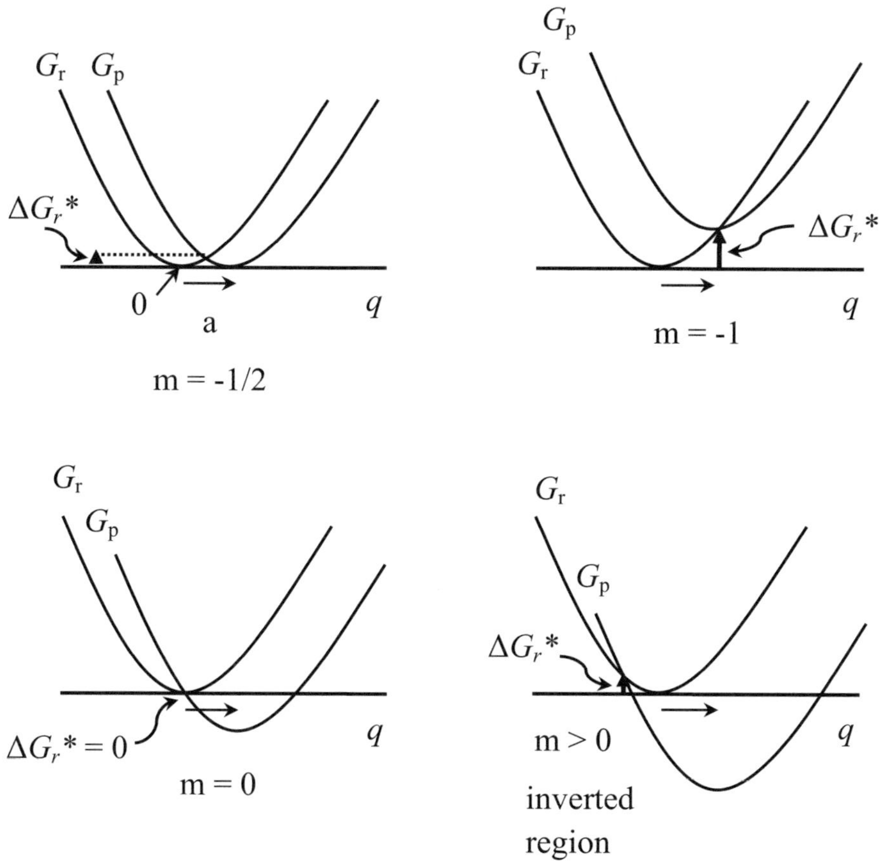


**Fig. 3.10** This shows four plots of $G_r$ and $G_p$ energies as functions of a generalized reaction coordinate, $q$, for different m values. The activation energies are $\Delta G_r* = G_p - G_r$. On each case the product has the position of its minimum increased by an amount a. When m = ½ the minima are the same energy and $\Delta G_r*$ is positive. When m = −1, the product curve moves up and $\Delta G_r*$ is positive. When m = 0 the product curve moves down and $\Delta G_r*$ is 0. When m is greater than 0 the product curve moves down further and $\Delta G_r*$ is again positive; this is the inverted region

## 3.7 Hush's Model Relating Transfer Coefficients and Electron Density Delocalization

Others were working on the problem of understanding electron transfer reactions with particular interest in the electrochemical interface. In 1958 N. S. Hush presented a theory for electron transfer reactions to or from ions near an electrode surface [15]. Hush expressed his model in terms of chemical potentials.

Let the Hush approach be expressed for a single electron reduction reaction of an ion in solution:

$$M^{n+}(aq) + e^{-}(s) \rightarrow M^{(n-1)+}(aq) \tag{3.25a}$$

He proposed a model showing the ionization potentials and electron affinities of the gas phase ions and the thermodynamic workfunctions of pure metals used for electrodes at equilibrium potential for electron transfer reactions *do not* determine the activation energies for ions in the double layer at the electrochemical interface. The independence of the activation energies from gas phase ionization potentials and electron affinities of the ions may not be obvious at first. At the time there was uncertainty about the effects of solvation of the ions though it could be expected that solvation would markedly shift their ionization potentials and electron affinities. In the Gurney model and Butler-Volmer theory the workfunctions of pure the metals, as reported for measurements in vacuum, were not parameters in the equations and so most likely the reported gas ionization potentials for the ions in the gas phase also should not appear in equilibrium and kinetic equations. If the electrode is a non-catalytic source or sink for electrons, reaction rates for electron transfer to non-adsorbed ions will only depend on the $E_f$ corresponding to the potential applied and the properties of the solvated ions. Hush started with the standard free energy of an electron in the metal, called the inner potential. The ions involved were solvated and were not adsorbed or otherwise influenced by the electrode beyond interacting with the double layer field. It is noted that the actual potential drop at an electrode–electrolyte interface is ill-defined because the ions and electrolyte molecules have positions and volumes meaning an accurate treatment requires determining the Gibbs energy of the interface with all of its components. Nevertheless, it is instructive to go through Hush's proof since it introduces definitions of some commonly used nomenclature. Hush carefully discussed the approximations he assumed in the modeling which may be seen in his paper. The ones essential to the terms in the formalism are given here.

Beginning with the double layer model, Fig. 3.11 shows a schematic for the double layer potential associated with charging an electrode surface positive of the pzc. In the equations that follow $\varphi_{sol}$ times the ion charge is used to represent the energy of the in the double layer field without specifying a position.

Hush chose reference states for the reactants as one mole of $M^{n+}$ ions plus one mole of electrons, and for the products one mole of $M^{(n-1)+}$, all in the gas phase at unit activity. The electron energy was assigned the vacuum level value, 0. Then Eq. (3.25a) is written for reaction in the gas phase as Eq. (3.25b):

$$M^{n+}(g) + e^{-}(g) \rightarrow M^{(n-1)+}(g) \tag{3.25b}$$

In this equation the standard Gibbs energy for the electron attachment was $\mu_I{}^0$. The standard electron affinity for the gas phase $M^{n+}$ ion was $-\mu_I{}^0$. The standard Gibbs energy of an electron in the pure metal was $\mu_e{}^0$. The negative of $\mu_e{}^0$ is the standard workfunction of the metal.

To describe the reaction when in solution, environmental components corresponding to Gibbs energies of hydration, $\Delta\mu_h{}^0(n)$ and $\Delta\mu_h{}^0(n-1)$ were added to $\mu_I{}^0$. An electrostatic potential term written simply as $\varphi_{sol}$ was added to take into

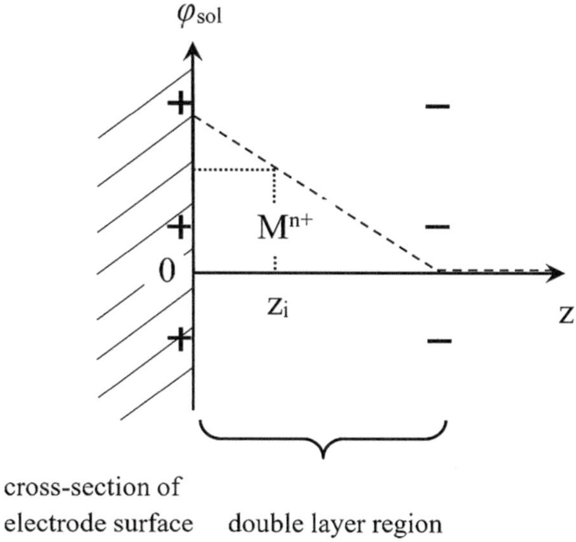


**Fig. 3.11** Cross-sectional view of an electrochemical interface showing an idealized double layer region with a positive surface charge and a negative compensating ion layer some distance away. The electric field decreases linearly and the potential experienced by a + 1 test charge at distance $z_i$ from the electrode surface is marked relative to its value in the bulk

account the environment in the double layer. The electron participating in the reaction required adjustments to its Gibbs energy due to its originating in the electrode material and responding to surface charging and interactions with the solution, and this was done by assigning it the inner potential $\varphi_M$. The inner potential of a metal is the work required to bring a point charge from infinity to a point beneath the surface of the metal. Though this is not measurable [16] it is interpreted as the adjustable surface potential for deriving formulas for symmetry factors and activation energies in Hush's model. In the bulk solution the reactants in Eq. (3.24) then have the Gibbs energy

$$\mu_r = \Delta\mu_h^0(M^{n+}) + n\varphi_{sol}F + \mu_e^0 - \varphi_M F \tag{3.26}$$

and the products have the Gibbs energy

$$\mu_p = \mu_I^0 + \Delta\mu_h^0\left(M^{(n-1)^+}\right) + (n-1)\varphi_{sol}F \tag{3.27}$$

The Gibbs reaction energy is then

$$\Delta G_r = \mu_I^0 + \Delta\mu_h^0(M^{(n-1)+}) - \Delta\mu_h^0(M^{n+}) - \mu_e^0 + (\varphi_M - \varphi_{sol})F \tag{3.28}$$

Equations (3.25)–(3.27) form may be expressed in Haber cycle form in Fig. 3.12.

There is a condition on the potentials: at equilibrium $\Delta G_r = 0$, which leads to the equation

$$(\varphi_M - \varphi_{sol})F = -\Delta\mu_h^0\left(M^{(n-1)^+}\right) + \Delta\mu_h^0(M^{n+}) - \mu_I^0 + \mu_e^0 \tag{3.29}$$

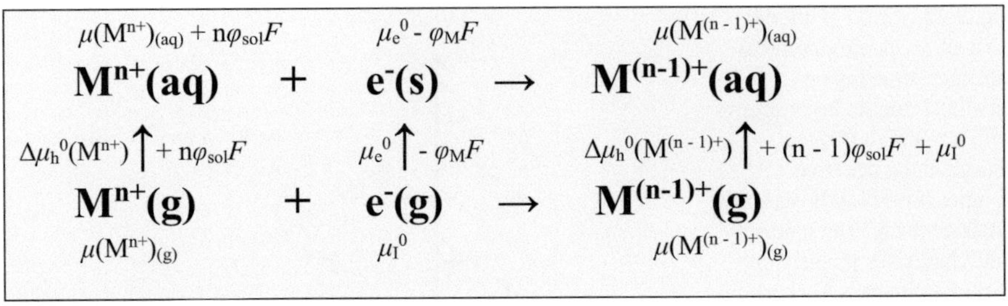


**Fig. 3.12** This Haber cycle-like figure takes the gas phase reactant $M^{n+} + e^-$ and product $M^{(n-1)+}$ to the aqueous phase reactants and products with upward arrows dressed with the energy changes of Eqs. (3.26) and (3.27), respectively. The result is that the Gibbs reaction energy given in Eq. (3.28) expressed visually in terms of the steps

Equation (3.29) contains the standard workfunction of the metal and the standard electron affinity of the gas phase ion but these will drop out when determining the activation energy for the electron transfer reaction, which is the energy difference between initial and transition state structures.

Now turning to the transition state for a single electron transfer from the electrode to the ion. Hush assumed that in the transition state the electron was delocalized over the metal and ion and metal wrote the fraction of the transferring electron on the ion in at the transition state as $\rho_I$* with

$$\rho_I^* = 1 - \rho_M^* \tag{3.30}$$

where $\rho_M$* is the fraction of charge on the metal in the transition state. The total charge on the ion in the transition state, $\rho^\ddagger$ is then

$$\rho^\ddagger = n - \rho_M^* \tag{3.31}$$

The Gibbs energy of the transition state is then

$$\begin{aligned}\mu_t^0 &= \left(n - \rho^\ddagger\right)\mu_I^0 + \Delta\mu_h^0\left(M^{(\rho^\ddagger)^+}\right) \\ &+\rho^\ddagger\varphi_{sol}F + \left(1 - n + \rho^\ddagger\right)\mu_e^0 - \left(1 - n + \rho^\ddagger\right)\varphi_M F\end{aligned} \tag{3.32}$$

and the activation Gibbs energy for the forward reaction becomes

$$\begin{aligned}\Delta G_t^* &= \left(n - \rho^\ddagger\right)\mu_I^0 + \Delta\mu_h^0\left(M^{(\rho^\ddagger)+}\right) \\ &-\Delta\mu_h^0\left(M^{n^+}\right) - \left(n - \rho^\ddagger\right)\mu_e^0 + \left(n - \rho^\ddagger\right)(\varphi_M - \varphi_{sol})F\end{aligned} \tag{3.33}$$

Substituting for $(\varphi_s - \varphi_0)F$ from Eq. (3.29), this becomes, under the equilibrium condition,

$$\Delta G_t^* = \Delta\mu_h^0\left(M^{(\rho^\ddagger)+}\right) - \Delta\mu_h^0(M^{n+}) + (n - \rho^\ddagger)\left[\Delta\mu_h^0(M^{n+}) - \Delta\mu_h^0(M^{(n-1)+})\right] \tag{3.34}$$

This activation energy depends only on the solvation energies of reactant and product ions, the charge of the transition state structure, and the solvation energy of the transition state complex. The thermodynamic workfunction of the pure metal, $-\mu_e{}^0$, is not present in the activation energy for the non-catalyzed reaction. Furthermore, the only properties of the reactant, transition state, and product ions present in Eq. (3.34) are hydration energies and not any property of the ions, including their gas phase ionization potentials and electron affinities.

The activation energy of Eq. (3.33) is put into Eq. (3.8) for reduction current density, where the transfer coefficient $a_c$ is factor multiplying the overpotential, $\eta$. The potential-dependent term in Eq. (3.33) is $(n-\rho^\dagger)(\varphi_M-\varphi_{sol})F$ so the cathodic current density has the form

$$i_c = \text{Constant}\left[\exp(-\alpha_c F\eta/RT)\right] \tag{3.35}$$

where

$$\eta = (\varphi_M - \varphi_{sol}) \tag{3.36}$$

and

$$\alpha_c = \left(n - \rho^\ddagger\right) \tag{3.37}$$

Hush derived a formula for the activation enthalpy based on detailed consideration of terms contributing to it, many of which were taken to be negligible. The major contributing term was an electrostatic energy that was estimated using the Born charging equation. The result is

$$\rho^\ddagger = (n - 1)/2 \tag{3.38}$$

In cases of symmetrical activation energy barriers $\alpha_a = \alpha_c = 1/2$. Further details for other reaction situations and comparison with experiment are not given here but maybe found in Hush's paper.

As shown earlier, in the Gurney approach to electron transfer at the electrochemical interface $\alpha$ depends on the ratio of slopes of reactant and product activation energies as functions of the separation distance. The derivation used the Franck–Condon approximation and the transition state was one of equilibrium between reactant and product with no partial transfer of the electron present in the description. In Hush's model for outer-sphere adiabatic electron transfer, partial electron transfer determined the activation barrier and the transfer coefficient. Marcus also provided an expression for the transfer coefficient and proposed that it should be close to ½ [14]. An ab initio quantum mechanical computational approach to finding $\Delta G_t{}^*$

will be presented in a later chapter and $\alpha$ will be potential-dependent and, like the Gurney theory, not associated with fractional charge [17]. A preview of the finding may be seen by going back to Fig. 3.6, where the slopes of the activation energies are $\alpha(U)$. The more complete quantum mechanical computational treatment, which is undergoing continuing improvement, is outlined in a later chapter. Surface charging and electrolyte polarization are included in the modern theory, making the electrode potential adjustable, and partial charge transfer is predicted at the transition states of adiabatic electrochemical reactions on electrode surfaces.

## References

1. Libby WF (1952) Theory of electron exchange reactions in aqueous solution. J Phys Chem 56:863–868
2. Debye P, Hückel E (1923) Zur Theorie der Elektrolyte. I. Gefrierpunktserniedrigung und verwandte Erscheinungen. Physikalische Zeitschrift 24:185–205
3. Debye P, Hückel E (1954) On the theory of electrolytes. I. Freezing point depression and related phenomena. The Collected Works of Peter J. W. Debye, Interscience, New York, 217–263
4. Marcus RA (1993) Electron transfer reactions in chemistry: theory and experiment. Angew Chem Int Ed Eng 32:1111–1121
5. Marcus RA (1956) On the theory of oxidation-reduction reactions involving electron transfer. I. J Chem Phys 24:966–978
6. Marcus RA (1956) Electrostatic free energy and other properties of states having nonequilibrium polarization I. J Chem Phys 24:979–989
7. Marcus RA (1957) On the theory of oxidation-reduction reactions involving electron transfer. II. Applications to data on the rates of isotopic exchange reactions. J Chem Phys 26:867–871
8. Marcus RA (1960) Exchange reactions and electron transfer reactions including isotope exchange, theory of oxidation-reduction reactions involving electron transfer. Part 4. A statistical-mechanical basis for treating contributions from solvent, ligands, and inert salt. Disc Faraday Soc 29:21–31
9. George P, Griffith JS (1959) Electron transfer and enzyme catalysis. Enzymes, Chapt. 8:347–389
10. Marcus RA (1960) Exchange reactions and electron transfer reactions including isotope exchange, theory of oxidation-reduction reactions involving electron transfer. Part 4. A statistical-mechanical basis for treating contributions from solvent, ligands, and inert salt. Disc Faraday Soc 29:21–31
11. Marcus RA (1963) On the theory of oxidation-reduction reactions involving electron transfer. V. Comparison and properties of electrochemical and chemical rate constants. J Phys Chem 67:853–857
12. Closs GL, Calcaterra LT, Green NJ, Penfield KW, Miller JR (1986) Distance, stereoelectronic effects, and the Marcus inverted region in intramolecular electron transfer in organic radical anions. J Phys Chem 90:3673–3683
13. Marcus RA (1965) On the theory of electron-transfer reactions. VI. Unified treatment for homogeneous and electrode reactions. J Chem Phys 43:679–701
14. Marcus RA, Sutin N (1985) Electron transfers in chemistry and biology. Biochim Et Biophys Acta 811:265–322
15. Hush NS (1958) Adiabatic rate processes at electrodes. I. Energy-charge relationships. J Chem Phys 28:962–972

16. Santos E, Schmickler W (2024) The inner potential in electrochemistry. J Solid State Electrochem 28:1319–1322
17. Anderson AB, Sidik RA, Narayanasamy J, Shiller P (2003) Theoretical calculations of activation energies for Pt + $H^+$(aq) + $e^-(U) \rightleftharpoons$ Pt-H: activation energy-based symmetry factors in the Marcus normal and inverted regions. J Phys Chem B 107:4618–4623

# Chapter 4
# Further Explorations on the Gurney-Butler and Marcus-Hush Themes

**Abstract** The Gurney-Butler approach to electron transfer was expanded by Bockris, Dogonadze, Kuznetsov and Levich with a focus on symmetry factors. They concluded that the symmetry factor is ~1/2 at the reversible potential and trends to zero at high overpotentials. Savéant, Hammes-Schiffer and Stuchebrukhov, Koper, and Schmickler extended the Marcus picture to include Born–Oppenheimer bond potentials, bond breaking, solvent relaxation, and quantum mechanical treatment of electronic wavefunctions during proton transfer. The models were becoming sophisticated computationally but the dependencies of polarization of the double layer, polarization of adsorbate bonds, and adsorbate structures on applied potential were not yet incorporated into the theory.

## 4.1 General Introduction

The successes of the Gurney model in its use of electronic structures to explain electron transfer and the Marcus theory and its applications to electron transfer reactions involving bond breaking inspired exploration by many other researchers. The results are significant enhancements to the model but do not include surface charging and electrolyte polarization on which cause potential-dependent variations in bond properties at the electrochemical interface.

## 4.2 Contributions to the Gurney Theory by Bockris and Matthews, Dogonadze, Kuznetsov, and Levich

The Gurney theory for electron transfer reactions applies to adiabatic conditions either with or without the breaking and making of chemical bonds, whether inner sphere or outer sphere. The Horiuti-Polanyi additions are the same as Butler's. There have been several later works following up on the Gurney-Butler theory since the papers were publication in the 1930's.

 
A. Anderson, *Past and Current Models in Electrochemistry*,
https://doi.org/10.1007/978-3-031-97791-6_4

Bockris and Matthews reformulated the Parson and Bockris study fifteen years later, [1] and developed formulas similar to those of Gurney and identified electron transfer activation energy by using Born–Oppenheimer potentials in the same manner as the Parsons-Bockris paper. They assumed equal distribution of the electron between the oxidized and reduced states, that is, on average the electron distribution would be ½ on the metal and ½ on the $H^+$ nucleus. The assumption of resonance was the justification. They reiterated the model of Gurney and Butler that the symmetry factor is a function of slopes of the reactant and product Born–Oppenheimer potential functions and are not dependent on charge transferred in the transition state.

Dogonadze, Kuznetsov, and Levich showed that reorganization of the solvent significantly contributes to activation energies for hydrogen ion discharge on metals [2]. These authors showed that in a range around the equilibrium potential the symmetry factor is around ½, and in the low overvoltage limit $\alpha$ becomes 1, and at high overpotentials $\alpha$ becomes 0.

## 4.3 Extending the Marcus Formalism: Work of Savéant, Hammes-Schiffer and Stuchebrukhov, Koper, and Schmickler

There have been significant efforts to extend the Marcus formalism and the understanding of outer-sphere electron transfer reactions. Savéant modified the Marcus model to include bond dissociation with electron transfer. He used intersecting Morse potentials for reactant and product and assumed adiabatic following of the electronic wavefunction during bond breaking [3]. To the activation energy component attributed to the intersection of the Morse potentials he added the solvent reorganization activation energy contribution and from several comparisons calculated activation Gibbs energies with experimental measurements determined bond breaking to contribute ~80%. Hammes-Schiffer and Stuchebrukhov included solvent relaxation in their formalisms for coupled outer-sphere electron and proton transfer reactions [4]. To see the progress with this viewpoint, a review by is available [5]. Koper [6]. used the Newns-Anderson Hamiltonian [7, 8] with additions by Schmickler [9] to discuss limiting cases. These included dependencies on high and low pH for reactions with electron transfer occurring after or before proton transfer as well as pH independent electron transfer reactions.

## 4.4 Summary of the Gurney, Marcus, and Hush Models and the Need for Powerful Computers to Progress Beyond Their Limitations

Three general theories for electron transfer have been introduced at this point. First is the theory of Gurney for ions over a range of distances from the electrode surface, extended by Butler, Horiuti and Polanyi, Bockris and coworkers to include effects of bonding to the electrode. A beauty of the Gurney theory for giving Tafel's linear log$i$ *vs* $U$ formula was is it did not require detailed density of state functions for electrons in metals or ion energies in solution because the Fermi distribution function was used for the available electron energy states and the Boltzmann distribution was used for the available ion energy states. Because of this, for a reduction reaction by a single electron log$i$ was to good approximation proportional to a constant plus $\eta$, corresponding to linear Tafel plots. Gurney and followers showed how to use bond stretching potential functions to calculate estimates of the transfer coefficient. Second is the theory of Marcus for outer-sphere electron transfers, extended by many workers, some of whom are mentioned above. These works employed harmonic vibrational potentials for coordination shell solvent molecules with solvent orientation energy added. Hush treated adiabatic electron transfer with reactant, transition state, and product charged and determined energies in the presence of a dielectric. In this model, the transfer coefficient equaled the reactant charge when at the transition state. For further discussion of these and related topics up through 1999 the author recommends consulting a book by Kuznetsov and Ulstrup [10].

Missing from the models discussed above are the effects of charging the electrode surface which will induce concomitant (i) polarization of the double layer, (ii) changes in adsorbate reactant and product structures, and (iii) shifting of $E_{\mathrm{f}}$. Predictions incorporating these features are beyond the reach of the electrostatic models and require quantum mechanical calculations. The slowness in moving in this direction after the development of the models of Gurney, Butler, Marcus, and others was largely a consequence of the lack of sufficient computing power to accurately handle large molecules and surfaces of metals. The earliest applications of quantum mechanical calculations with the goal of understanding and predicting the effects of charges applied to an electrode focused on spectroscopically measurable potential-dependent properties of adsorbates. These were approximate parameterized semiempirical calculations of wavefunctions and energies for models of the electrode surface and adsorbates bonded to them. Perturbation theory made use of energy levels and orbital wavefunctions produced by the calculations to provide interpretations of the observed potential dependencies.

## References

1. Bockris JO'M, Matthews DB (1966) The mechanism of charge transfer at electrodes. Proc Roy Soc Lond, Math Phys Sci 292:479–488
2. Dogonadze RR, Kuznetsov AM, Levich VG (1968) Theory of hydrogen-ion discharge on metals: case of high overvoltages. Electrochim Acta 13:1025–1044
3. Savéant J-M (1987) A simple model for the kinetics of dissociative electron transfer in polar solvents. Application to the homogeneous and heterogeneous reduction of alkyl halides. J Am Chem Soc 109:6788–6795
4. Hammes-Schiffer S, Stuchebrukhov AA (2010) Theory of coupled electron and proton transfer reactions. Chem Rev 110:6939–6960
5. Warburton RE, Soudackov AV, Hammes-Schiffer S (2022) Theoretical modeling of electrochemical proton-coupled electron transfer. Chem Rev 122:10599–10650
6. Koper MT (2013) Theory of the transition from sequential to concerted electrochemical proton-electron transfer. Phys Chem Chem Phys 15:1399–1407
7. Anderson PW (1961) Localized magnetic states in metals. Phys Rev 124:41–53
8. Newns DM (1969) Self-consistent model of hydrogen chemisorption. Phys Rev 178:1123–1135
9. Grimminger J, Bartenschlager S, Schmickler W (2005) A model for combined electron and proton transfer in electrochemical systems. Chem Phys Lett 416:316–320
10. Kuznetsov AM, Ulstrup J (1999) Electron transfer in chemistry and biology: an introduction to the theory. John Wiley & Sons, Ltd

# Chapter 5
# Interplay of New Spectroscopies and Semiempirical Quantum Theory Applied to Electrochemistry: 1970s–1980s. The Chemical Bond is Robust Under Perturbation

**Abstract** In the 1970s and 1980s spectroscopic probes of surfaces interfaced with vacuum were developed. They revealed adsorbate surface structure patterns and electronic and vibrational for the adsorbed molecules and even their structures. Some surface sensitive spectroscopies were applicable to electrochemical interfaces. When the electrode potential is changed the surface electron chemical potential ($\mu_e$) and Fermi energy ($E_f$) change proportionally. In cases with robust charged surface and double layer, emersed electrodes interrogated with UPS and XPS demonstrated the proportionality. Understanding the electrochemical vibrational tuning rate (EVTR) of adsorbed molecules such as CO spurred the development of two theoretical viewpoints, one focusing on the electron chemical potential and valence electron band position of the electrode in the sense of Gurney and the other focusing on the presumed double layer electric field. Perturbation theory and semiempirical MO theory that could be applied to understand EVTR trends with the Gurney picture are outlined. Theories relating EVTR to bond lengths and force constants are reviewed. To provide an ability to calculate EVTR trends, a theory relating the vibrational force constants to bond lengths using electron density distribution functions in atoms is also outlined. Application of the electric field Stark effect as a model to explain EVTR is discussed.

## 5.1 General Introduction

Developments outlined so far are mostly general models of the electrochemical interface and make limited use of bonding and other properties of molecules beyond hydrogen. The focus is on kinetics based on Born–Oppenheimer potential energy functions for atom interactions and on an electronic structure model for the exchange of electrons between ions and between molecules and external electron sources or sinks of chosen potential. Surface sensitive spectroscopies advanced in the 1970's and 1980's began delivering details for adsorbed molecules such as vibrational frequencies. These included low energy electron diffraction (LEED), ultraviolet and x-ray photoemission spectroscopies (UPS and XPS), high resolution electron energy

 
A. Anderson, *Past and Current Models in Electrochemistry*,
https://doi.org/10.1007/978-3-031-97791-6_5

loss spectroscopy (HREELS), Fourier transform infrared reflection absorption spectroscopy (FT-IRAS), surface-extended X-ray absorption fine structure spectroscopy (SEXAFS), surface-enhanced Raman spectroscopy (SERS), scanning tunneling microscopy (STM) and sum-frequency generation spectroscopy (SFG). On electrode surfaces vibrational excitations could vary with the applied potential. For example, measured spectra of vibrational absorptions for CO adsorbed on Pt(111) electrodes had well-defined intensity groups representing different adsorption states and their intensities varied with the applied potential meaning the occupation of each state varied with the applied potential. Not only that, the frequencies of peaks in the groups also changed with potential. The frequencies trended to lower values as the potential decreased, meaning as the electron chemical potential or $E_f$ shifted up in the energy scale the CO bond became weaker. The adsorption states in the CO on Pt(111) example corresponded to 1-, 2-, and threefold coordination. What caused dependencies of vibrational frequencies on the site of adsorption and what caused site occupations and frequencies to depend on the applied potential? How should we think about such modified CO?

The concept of bonds was of course known to chemists and physicists and most had a copy of Pauling's famous book on their shelf for reference on chemical bonds and structures [1]. With the evolving spectroscopies able to give information about what was on electrode surfaces, and spurred by the energy crisis of the 1970's, funding agencies began to strongly support surface science and electrocatalysis related research with a view toward developing fuel cell batteries for transportation, portable power, and stationary power applications. Unraveling the complicated chemistry with a view toward enhancing the efficiency of the four-electron electroreduction of $O_2(g)$ to $2H_2O(l)$ was one of the main themes and provides a focus for much of what follows later in this book. Reducing oxygen to water is the cathode reaction in hydrogen fuel cells and this reaction suffers a much greater energy loss than the hydrogen oxidation to $H^+(aq)$ at the anode. Molecules such as adventitious CO and intermediates in $O_2$ reduction such as OH were still viewed as molecules when bonded to a surface. From the above-mentioned CO on Pt(111) example, adsorbed molecules were expected to be robust, meaning they should have much the same structural, spectroscopic, and chemical reactivity characteristics as when in the gas or solution phases. When perturbed by changes in the applied potential they would then retain recognizable characteristics up to the point of bond breaking or Faradaic reaction. Following bond dissociation, reaction with other molecules either in solution or on the surface might follow. A Faradaic reaction might include, for example, protonation. Bond strengths have long been assumed by chemists to be the key to understanding non-electrochemical catalysis [2]. The procedure for catalysis prediction from bond strengths would be through Hammond's postulate [3]. Given the relationship between reaction energies and reversible potentials for reducing molecules in solution, it follows that when the reactant and/or product are adsorbed on an electrode the reaction energy will be different, making the reversible potential different. Then it is possible to start with known standard reversible potentials from experimental tabulations and predict reversible potentials for the same reaction as intermediates on the surface of an electrocatalyst. In other words, adsorption bond strengths perturb the energy and reversible potential of the bulk solution phase reaction to give the catalytically relevant reversible potentials for reactants and products when one or both are adsorbed on the electrocatalyst. A later chapter will develop this perturbation theory concept, which in implementation is called the linear Gibbs energy relation (LGER). The LGER model will be compared and contrasted with the well-known computational hydrogen electrode (CHE) model, which instead of using experimentally determined standard reversible potentials for input data, effectively calculates them when making predictions.

In electrochemistry a molecule adsorbed on an electrode is perturbed by the bond it forms with the surface, which is itself perturbed not only by changes in the electrode potential and Fermi energy, but also by the surrounding electrolyte and ions or molecules contained in the electrolyte that participate in reaction of the adsorbed molecule. The Fermi energy level moves when the potential being applied is changed and electric fields near the surface of the electrode also change. The fields have often been the variable of greater attention in electrochemistry. The concept of double layer field comes from electrolyte polarization in response to surface charging. The result of the polarization is at the simplest imagined to be a plane of oppositely charged ions above the charged surface, the whole arrangement being a capacitor. The resulting electric field between the plates being the double layer field. Chemists often think about chemical bonding perturbations caused by charging the surface and physicists often think about adsorbed molecules immersed in electric fields. Both views were used to attempt explanations for variation in CO vibrational frequencies when adsorbed on Pt(111) electrodes when the applied potential was changed. This chapter provides a detailed analysis of EVTR models based both on $E_{\mathrm{f}}$ shifts and model electric fields and stresses the point that electric fields are derived from the charge density distribution functions at the interface. If the quantum mechanical Hamiltonian can be solved to give these functions, at the same time the it can yield vibrational frequencies from the total energies without reference to the electric fields, though they be calculable.

A molecule dissolved in an electrolyte experiences interactions with the surrounding molecules and ions and these may include solvation stabilization, i.e., the coordination of solvent molecules around the solute, steric interaction energies, and perhaps electric field influences from nearby solvated ions. The Marcus and Hush theories approximated the solvent component. For molecules with relatively strong bonds none of these interactions are disruptive but they cause shifts in vibrational and electronic spectra from the values of the molecules in the gas phase.

Bonding interactions of molecules with an electrode surface are associated with vibrational properties and also with electrochemistry. The LGER method, which relates potentials for reactions on surfaces to their bulk values via adsorption energies, is an example that, along with others, will be given detailed treatment in later chapters. Bonding of electrochemical reaction intermediates such as CO, OH, or OOH to the surface of a catalyst such as platinum is relatively strong. The OH and OOH radicals form strong σ bonds to the surface. The CO molecule bonds strongly to many open shell d-band metals because of mixing of its π and σ MOs with metal s- and d-band orbitals so that there is a combination of σ and π bonds between CO and the surface.

Pauling used covalent bond theory, the octet rule, and resonance to explain chemical bonding. A different viewpoint lies with molecular orbital (MO) theory, where the electron wavefunctions are delocalized over a molecule or atom cluster and are well approximated using various MO computational codes. Principles of MO theory are introduced in this chapter and they provide the interpretive approach to chemical bonding used in this book.

## 5.2 Electron Chemical Potential and Workfunction Depend Linearly on Applied Potential

Quantum chemical calculations of potential dependencies of adsorbate properties began about the same time that spectroscopic measurements on emersed electrodes showed that electrode surface workfunctions depended on applied potential in 1:1 proportion. This is seen in the experimental studies initiated in 1978 by Hansen and coworkers [4]. They refined their work in a series of papers and others followed with measurements for a variety of electrode materials [5]. The studies made it clear that the surface electron chemical potential and workfunction varied 1:1 with changes in electrode potential measured in volts. Figure 5.1 from Kötz et al. [5] shows how the relation holds over different metals, Ag, Au and Pt and Fig. 5.2 and how it holds over Pt even as the potential increases through surface oxide formation. It means that in aqueous electrolytes the electron chemical potentials at an electrode surface can generally be expected to be tied to the applied potential. That this is the case means the electrode on emersion has attached to it a double layer at the potentials of the measurements. The workfunction measurements by electron emission processes required the ejected electrons to pass through this double layer. If upon ejection from the surface the electrons interacted elastically with the double layer molecules the measurements would directly give the surface workfunction. Inelastic collisions would cause some of the ejected electrons to have lower energies and this would serve to broaden and possibly shift the emission edge. In the case of the oxidized electrode surfaces the ejected electrons will pass through the oxide layer as well as the electrolyte double layer. Figure 5.2 shows linear change in workfunction as the potential is increased through the oxide formation range. Assuming the oxide coverage is complete, this means the collisions experienced by ejected electrons when passing through the oxide as well as when passing through the double layers on the metal or on the oxide are for the most part elastic.

The relationship between potential and electrode workfunction is then

$$eU = \varphi + \text{const} \tag{5.1}$$

Based on the emersion studies, Eq. (5.1) holds for electrode surfaces in general, no matter the metal composition or the electrolyte used. This result opens the door to quantum mechanical modeling of responses in adsorbed molecules to changing the

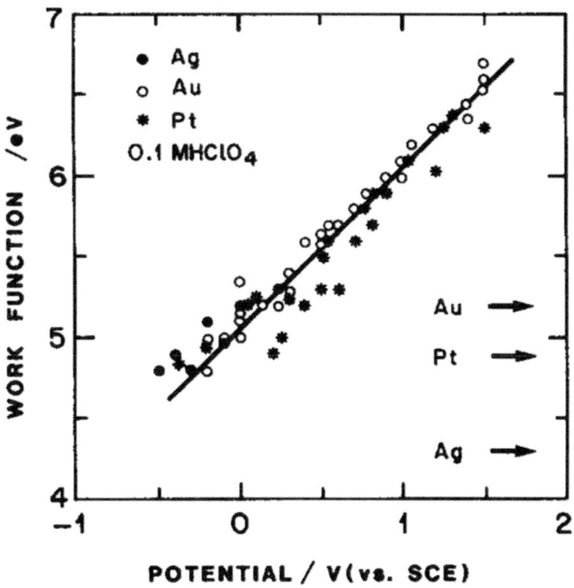


**Fig. 5.1** A graph of numerous ex-situ measurements by Kötz et al. of electrode workfunctions versus potential for gold, platinum and silver all lie near a trend line with slope near 1. Arrows indicating workfunctions of clean metal surfaces. Reprinted from Journal of Electroanalytical Chemistry and Interfacial Electrochemistry, Vol 215, E.R. Kötz, H. Neff, K. Müller, A UPS, XPS, and Workfunction Study of Emersed Silver Platinum and Gold Electrodes, 331–344, Copyright 1986, with permission from Elsevier

workfunction and $E_f$ by using semiempirical MO theories parameterized with atom valence state ionization potentials (VSIP). Through them the electrode's valence band and workfunction are controlled, as will be presented later. Gurney used this electronic structure framework to explain electron transfers to or from molecules that were not adsorbed. Then Butler added potential energy of bonding of the reactant to the electrode surface. These early studies and the previously introduced extensions of them assumed interaction energies were obtained satisfactorily by using rigid Born–Oppenheimer potential functions, such as Morse potentials, for the reacting bonds and shifting the reductant curve based on the change in the electron's energy at different potentials. Hush introduced electron delocalization for adiabatic electron transfer using electrostatics, an idea supported later by advanced quantum mechanical methods, which are discussed later Chap. 9, Sect. 9.5.

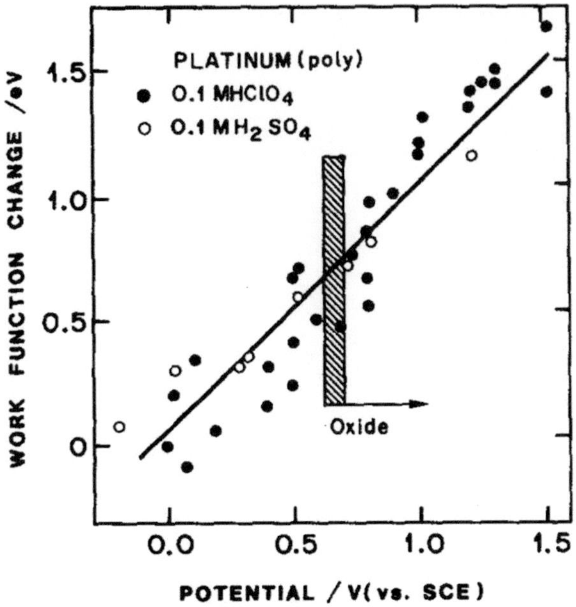


**Fig. 5.2** Electrode workfunctions versus potential measurements for platinum in tenth molar perchloric acid and tenth molar sulfuric acid from Kötz et al. Above about 0.6 V the surface becomes covered by an oxide layer but both sets of measurements are close to a line of slope 1 from zero to 1.5 V. Reprinted from Journal of Electroanalytical Chemistry and Interfacial Electrochemistry, Vol 215, E.R. Kötz, H. Neff, K. Müller, A UPS, XPS, and Workfunction Study of Emersed Silver Platinum and Gold Electrodes, 331–344, Copyright 1986, with permission from Elsevier

## 5.3 New Surface Sensitive Spectroscopies Revealed Structure, Vibrational, and Electronic Properties of Adsorbed Molecules

Spectroscopies for measuring structures and vibrational and electronic properties of surfaces and adsorbates at the vacuum interface were advanced beginning in the 1970s. For example, low energy electron diffraction (LEED) gave surface structure patterns; ultraviolet photoelectron spectroscopy (UPS) gave surface valence electron densities of states; x-ray photoelectron electron (XPS) gave surface core valence electron densities of states; high resolution electron energy loss spectroscopy (HREELS) and Fourier transform infrared reflection adsorption spectroscopy (FT-IRAS) gave surface bond vibrational energies; surface-extended X-ray absorption fine structure (SEXAFS) spectroscopy gave adsorbed coordination structures and states of molecules; scanning electron tunneling microscopy (STM) gave surface structures; surface enhanced Raman (SER) and sum frequency generation (SFG) spectroscopies gave vibrational energies for adsorbed molecules.

## 5.4 Stark Tuning Rates of Molecules in Electric Fields

A connection between the shifting $E_f$ in electrochemistry and a potential-dependent spectroscopic measurement was found in measurements of $d\omega/dU$, the rate of change in bond vibrational frequency, $\omega$, with change in applied potential, $U$, in molecules adsorbed on electrodes. These will be called electrochemical vibrational tuning rates (EVTR). Note that some papers use the symbol $\omega$ for frequency and others use $\upsilon$. However, $\omega$ is reported in units of wavenumbers, $cm^{-1}$ while $\upsilon$ is reported in Hertz. The energy change between two vibrational states is $\Delta E = h\upsilon = \hbar\omega$ where $\omega = 2\pi\upsilon$. Here h is Planck's constant and ħ is called h-bar and equals h divided by $2\pi$. In the following the symbol used for frequency will be that employed in the literature paper under discussion.

The first $d\omega/dU$ measurements for electrode surfaces did not refer to electric fields and it is not clear how they are related to the historical vibrational Stark effect with tuning rates which varied with electric fields to give $d\omega/dF$. The focus of many has been on electric fields when seeking explanations of EVTR. Early work by Bagus and Pacchioni modeled electrode surfaces with small clusters and introduced a single adsorbing atom or molecule bathed the systems in external electric fields and calculated vibrational shifts [6]. While the results are interesting, as shown below it is difficult to accurately determine the electric field at an electrode surface and it changes with the applied potential. Calculating the effects of $E_f$ shifts on small molecules bonded to the surface is relatively easy. This can be done using semiempirical MO calculations to shift $E_f$ without adding an electric field. Ideally, ab initio calculations including the polarizable electrolyte with double layer ions and adding charge to the surface to shift it could be performed to give more accurate EVTR. In the latter, the electric field would be a complicated but calculable function of position and structure but it is not needed for determining $d\omega/dU$.

The Stark effect for free molecules, discovered by J. Stark, measures the splitting of degenerate electronic state energies by electric fields [7]. E. U. Condon noted that there would would be shifts in molecular vibrational frequencies [8]. D. Bishop in 1993 summarized the developments over the span of years following Condon's suggestion [9]. He reviewed finite field quantum calculations of the vibrational Stark effect for free molecules, and presented a perturbation theory formalism for the effect, which is the change in vibrational frequency with respect to change in electric field, $F$, $d\omega/dF$. In it, the leading term in a Taylor series expansion of $\hbar\omega$ for a bond in an electric field and contains a factor equal to the difference in dipoles for the initial and final states, and the dipole difference is called the Stark tuning rate (STR).

## 5.5 Should EVTR Be Considered as Functions of Shifts in $E_f$ or as Functions of Assumed Electric Fields?

For simple interpretations of EVTR for electrochemical systems, two quantum computational model approaches can be pictured: (i) find or assume an electric field for each applied potential and use this to calculate the tuning rate and (ii) determine how the shift in $E_f$ and the valence electronic band that occurs when the potential is changed affects the bonding of the adsorbate to the electrode and how it affects the bonds within the adsorbate to give the tuning rates. The first can be accomplished by adding electric fields to the adsorbate–surface model Hamiltonian in the sense that Bishop did for free molecules and Bagus and Pacchioni had already done for molecules bonded to a cluster of metal atoms. More will be said about this later but there two warnings to keep in mind: (i) the field's magnitude and spatial dependence are not generally known and (ii) the electric field is a derived quantity from having already obtained wavefunctions and charge density distributions by solving the Schödinger equation for the interfacial system. If the accurate calculations of the field are accomplished, then the calculations also provide the Born–Oppenheimer potential energies for the determining the vibrational frequencies. Since adding and subtracting electron charge to the electrode surface changes the workfunction of the electrode surface and, according to Eq. (5.1) changes its potential, the potential dependencies of adsorbate vibrational frequencies are available from frequencies that can be calculated using the Schrödinger equation without explicit consideration of electric field.

Semiempirical MO theories can model valence band shifts that occur in response to surface charging and provide variationally optimized wavefunctions and energies in response to changes in applied potential. This allows approximate predictions of potential-dependent $\omega$. Wavefunctions from the same calculations might also be used to generate charge density distributions that could be used for calculating the surface electric field, but in the absence of surface charging the field function would be nearly constant and not productive for use in determining the EVTR. Valence band shift calculations build on the model of Gurney and extensions by Butler, Volmer, and later workers as presented in Chap. 1, and the model became the basis for understanding charge transfer kinetics in electrochemistry. It is therefore expected that the band shift approach using simple semiempirical quantum calculations of adsorbate–surface models will be useful.

In an ab initio treatment of EVTR, $E_f$ shifts because of the added charge and this charge is compensated by the polarizable double layer so that there is no net charge in the electrochemical interface. Such calculations explain EVTR in terms of potential-dependent changes in bonding; the electric fields which, though calculable using the variationally-determined wavefunctions, are not used. The semiempirical band shift model mimics the ab initio calculations of the potential dependence of bonding of adsorbed molecules but maintains the simplicity of omitting the double layer.

The remaining sections of this chapter develop the conceptual and computational tools used to address the ways EVTR is modeled using quantum theory. The last section considers Stark and EVTR tuning rates of moieties of molecules tethered to an electrode.

## 5.6 Brief Summary of ab Initio Hartree–Fock MO Formulas

The ab initio Hartree–Fock (HF) theory provides the framework for semiempirical MO calculating codes used in Extended Hückel (EH), atom superposition and electron delocalization (ASED), and others. Only key equations are given here. The details of the derivation are found most in text books on quantum theory. In the HF approach an atom or molecule with an even number, n, of electrons has the single determinant wave function

$$\Psi_{el}(1,\ldots,n) = (n!)^{-1/2}\left|\Psi_1(n)\alpha(n)\ldots\Psi_{n/2}(n)\beta(n)\right| \tag{5.2}$$

where $\Psi_i$ in the determinant are atomic orbitals (AOs) in the case of atoms or MOs in the case of molecules and α and β are the spin states available to an electron. Omitting electron repulsions, the sum of kinetic and potential energies obtained from solving the Schrödinger equation using this wavefunction is

$$E = 2\sum E_i \tag{5.3}$$

with the sum from 1 to n/2 and

$$E_i = \left\langle \Psi_i(1)\middle| -1/2\nabla^2(1) + V(1)\Psi_i(1)\middle|\right\rangle \tag{5.4}$$

In Eq. (5.4) atomic units (au) are assumed and $\nabla^2(1)$ is the Laplacian operator for electron 1 and $V(1)$ stands for all contributions to the electrostatic potential that is experienced by electron 1, including nuclear and other potentials external to the molecule, but excluding potentials from other electrons. When all electron repulsions are added the Hartree–Fock orbital energy emerges:

$$\varepsilon_i = E_i + \sum\left(2J_{ij} - K_{ij}\right) \tag{5.5}$$

where the sum is over j = 1 to n/2 and J and K are, respectively called coulomb and exchange integrals and the take into account the pairwise electron repulsion potentials. The Hartree–Fock total electron energy is for the closed shell system a sum of orbital energies multiplied by 2 from which the electron repulsion energy is subtracted because it is counted twice in the sum:

$$E_{HF} = 2\sum \varepsilon_i - \sum \left(2J_{ij} - K_{ij}\right) \tag{5.6}$$

Here the first sum is over i from 1 to n/2 the second sum is over i and j from 1 to n/2. Equation (5.6) gives the HF electronic energy and total molecular energy includes a pairwise sum over nuclear repulsion energies. A Fock operator F(1) can be defined so that

$$< \Psi_i(1)|F(1)|\Psi_i(1) >= \varepsilon_i \tag{5.7}$$

This equation invites semiempirical modeling as in EH theory in which parameters based on atomic properties are employed and it is not necessary to calculate electron repulsion integrals. The negative of an orbital energy $\varepsilon_i$ is called the Koopman's ionization potential for orbital i and within the HF model it equals the energy difference between the original system and the positive system with an electron removed and at zero energy [10]. In the case of molecules, linear combinations of atomic orbital MO (LCAOMO) basis functions are generally used. MO $\Psi_i(1)$ is the sum over the AOs:

$$\Psi_i(1) = \sum a_{i\mu}\varphi_\mu(1) \tag{5.8}$$

In Eq. (5.8) the a are coefficients and the φ are AOs. The (1) stand for coordinates $\boldsymbol{r}_1$ for one of the electrons.

Roothaan introduced a self-consistent-field (SCF) procedure based on linear variation theory for obtaining a coefficients and hence the MO'S and the MO energy levels [11]. When, for a particular molecule, the AO basis set contains m AO's the result of a Roothaan calculation for a closed-shell molecule with n electrons consists in m MO's and m energy levels with n/2 doubly occupied with opposite spin electrons and m—n/2 unoccupied. When displayed as functions of energy values for the $\varepsilon_i$, the result is the familiar electronic structure representation in terms of energy levels to which electron occupations are added. Looking at Eq. (5.6) we see that the total electronic energy is not the sum of electron orbital energies. Nevertheless, there is a long history of explaining chemical bonding changes in terms of one or more orbital energies. This ability means the electron repulsion term in Eq. (5.6), though not constant, often does not change the qualitative conclusions derivable from thinking about the calculated orbital energies and shapes.

## 5.7 Experimental Basis for Semiempirical Band Shifting Computational Models to Understand EVTR with MO Theory

The EVTR of CO adsorbed on transition metal surfaces has been a focus of theory for many decades. These studies provide excellent chalkboard examples for presenting the major developments in computational electrochemical methods that include the electrode potential as a variable. One of the earliest EVTR studies used surface enhanced Raman measurements of cyanide anion, which is isoelectronic with CO, and other nitrogen-containing compounds adsorbed on silver to measure potential dependence of the vibrational frequencies [12]. Another study employed reflectance infrared spectroscopy to measure the EVTR for the CO bond when adsorbed on platinum [13, 14]. It is noted that this early work referred to potential dependence and not electric field dependence.

## 5.8 Dewar-Chatt-Duncanson Bonding Picture

A theory of metal-adsorbate bonding already existed at the time of the early measurements of EVTR. It explains changes in vibrational frequencies of molecules like $CN^-$ and CO upon adsorption on uncharged surfaces. This is the Dewar-Chatt-Duncanson MO model for organometallic chemistry [15, 16]. In this theory for $\pi$-complexes, overlap of the empty antibonding ($\pi^*$) orbitals of the ligand molecule with the occupied *d* orbitals on the metal atom forms a metal–ligand bond and at the same time weakens $\pi$-bonding in ligands such as ethylene or CO.

## 5.9 Blyholder's Application to CO Adsorption

Blyholder applied the Dewar-Chatt-Duncanson bonding picture to CO adsorption on metal surfaces, but did not address electrodes with their variable potential. He considered adsorption sites (onefold and higher coordination bonding) and coverage [17]. The nature of the bonding was identified using Hückel MOs as $\pi$ interactions of CO with surfaces of Ni. For CO, *p* orbitals were assigned to each atom and for Ni atoms *d* orbitals were assigned. Standard Hückel matrix element parameters were used. More about Hamiltonians and quantum chemical calculations is given later, but it this point it is enough to know the essence of the Schrödinger wave equation solutions: pairs of AOs form two MOs a more energetically stable one with orbitals superimposed with no node and the less stable one with a node. Add a third AO and 3 MO's form, etc.

The result of the "3-state calculation" was 3 MOs with energy levels as in Fig. 5.3. This shows symmetry-allowed mixing of 2p orbitals on the C and O atoms with a

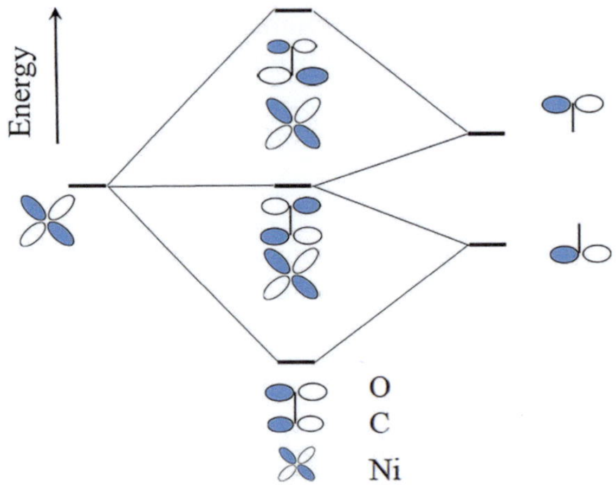


**Fig. 5.3** MO correlation diagram for CO bonded to a Ni atom based on the Blyholder calculation. Energy increases in the up direction. The Ni d orbital energy is the level on the left, and it lies between the higher-lying C p and lower-lying O p levels on the right. The column of levels in the center show a bonding Ni 4d + CO $\pi$ bonding MO at the bottom with no nodes, a non-bonding Ni 4d + CO $\pi$* MO in the middle with one node between p orbitals on C and O, and an antibonding Ni 4d—CO $\pi$* MO with three nodes at the top

4d orbital on Ni to give three $\pi$ orbitals centered on the three atoms. There is an identically shaped set of 4d and $\pi$ orbitals with the same energies perpendicular to this set, which are not shown. Sigma orbital formation along the linear NiCO axis, was not addressed in Blyholder's calculation. The two lowest orbitals in Fig. 5.3 are primarily on the CO, each occupied by two electrons. The middle two are antibonding between C and O and bonding between C and Ni, leading to lengthening of the CO bond because they are occupied, resulting in decreases the energies and frequencies of the CO vibrational states. The top two are antibonding throughout, high in energy and not occupied. Using this bonding model, Blyholder interpreted the influence of bulk Ni surface structures and CO coverages on the measured fundamental CO stretching frequencies. In his qualitative view, the contribution of the CO antibonding orbitals of the type of the middle one in Fig. 5.3 depended on the environment in which the molecule found itself when adsorbed.

## 5.10 Hoffmann's Extended Hückel MO Theory and Application to EVTR

At about the time Blyholder's paper was published, Hoffmann introduced the semiempirical extended Hückel (EH) MO theory which provided approximate valence MOs and energies for non-linear molecules [18]. These were useful for calculating bond angles in polyatomic molecules but did not give accurate equilibrium bond lengths. The theory follows the outline of Hartree–Fock theory but it omits nuclear repulsion energies and in place of the complete electron Hamiltonian it uses parameterized matrix elements:

$$H_{ij}^{ab}(\boldsymbol{R}_{\mathrm{ab}}) = (K/2)\left(H_{ii}^{a} + H_{jj}^{b}\right)S_{ij}^{ab}(\boldsymbol{R}_{\mathrm{ab}}) \tag{5.9}$$

Here the $H_{\mathrm{ii}}{}^{a}$ are approximately the energies of the valence AOs $i$ on atom $a$, $K$, called the Wolfberg-Helmholtz constant, is 1.75, $S_{ij}{}^{ab}$ is the overlap integral of orbital $i$ on atom $a$ with orbital $j$ on atom $b$, and $\boldsymbol{R}_{\mathrm{ab}}$ is the distance between nuclei of atoms a and b. The $H_{ii}$ contain at least part of electron repulsion energies within the atom. The $H_{ij}{}^{ab}$ matrix elements omit two-center electron repulsion contributions. The solution to this Hamiltonian is a set of MOs and their energies. The Hartree–Fock energy comes from solving the complete Hamiltonian using a Slater determinant of AOs and includes electron contributions from the MOs plus the nuclear repulsion energies minus the electron repulsion energies which are counted twice in the sum of MO energies. Hoffmann proposed that the latter two energies, being of opposite sign, approximately cancel, which helps explain the applicability of EH electronic energies to predicting and interpreting bond angles.

Applying EH calculations, Robertson and Wilmsen predicted CO was most stable when perpendicular with the C end in the twofold bridging site on Ni(100) [19]. This was confirmed about 20 years later when Grossmann, Erley, and Ibach deduced from FT-IRAS spectra that at low coverage CO bonded at the bridging site and for increasing coverage the onefold atop site also became occupied [20] The onefold sites became increasingly occupied upon increasing the temperature, which they explained with a lattice gas model.

Following the Robertson-Wilmsen EH calculations of the structure of CO on Ni(111) Anderson and Hoffmann calculated valence $\pi$ and $\sigma$ orbitals and energies for first row diatomic molecules and ethylene bonding to a clusters of Ni and W atoms representing (100) surfaces [21]. Consideration was made of $\pi$ and $\sigma$ orbital energy shifts occurring upon adsorption and accompanying charge transfers and bond weakening. Orbital sizes were seen to play a role and low coordinate edge and corner metal atoms gained electronic charge and became more attractive, possibly activating them, it was proposed, for dissociation of electrophilic molecules such as CO and ethylene.

Could EH calculations yield accurate EVTR values by shifting the metal valence band and Fermi energy in response to applied potential? It was proposed by Döhnert, Koutecky, and Schultze that electrode potential changes could be simulated in EH theory by shifting the surface metal atom ionization potentials [22] However, EH was not applied to calculating EVTR values because the potential energy curves needed to calculate bond vibrational force constants and frequencies were insufficiently accurate. An explanation of the trends in vibrational frequencies of adsorbate–surface bonds with changes electrode potential may be had from combining the MO interpretation of the bonding of CO to the electrode and the potential-dependent electronic band picture of Gurney. This is best understood using Hoffmann's PMO approach [23].

The Gurney picture of the electrode valence band and $E_f$ responding to changes in applied potential is in Fig. 5.4: as the potential is decreased, $E_f$ moves toward the empty adsorbate valence band and away from its filled donor band. The key result of perturbation MO theory is that when atoms and molecules come together and MOs form, the upper stabilizes the lower with bonding overlap and the lower destabilizes the upper with antibonding overlap. Figure 5.5 shows this for the $\pi$ and σ bonding interactions between Ni and adsorbed CO.

This figure differs from Fig. 5.3 in that the Ni d band showing $E_f$ is shown instead of a single d level and CO $\pi$ and $\pi$* molecular MOs are used instead of the AOs, and the resulting bonding and energy level description directly connect to measurements of photoelectron spectroscopy [24, 25]. The sigma donation from CO to Ni and back-donation from Ni to CO $\pi^*$ orbitals seen in Fig. 5.5 follows the Dewar-Chatt-Duncanson picture and the energies are based qualitatively on the photoelectron measurements of Turner.

Photoelectron spectroscopy provided molecular energy level structures with which the energy levels of MO calculations agreed, whereas energy levels calculated using localized covalent orbitals were not in agreement with the measured spectra.

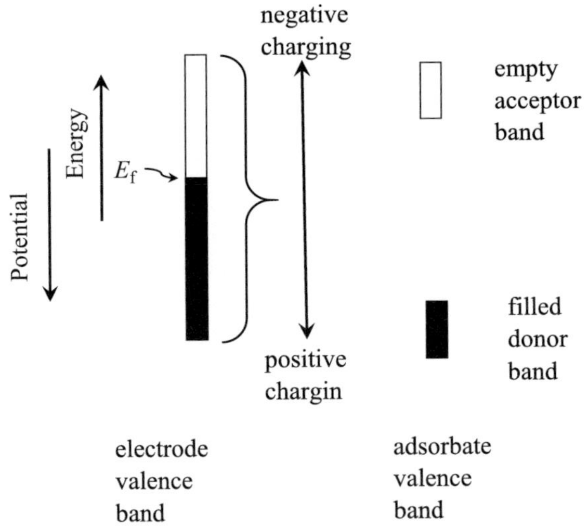


**Fig. 5.4** A figure showing on the left the electrode valence band, with energy increasing in the up direction, and $E_f$ marked near its middle. To the right are an empty adsorbate acceptor valence above $E_f$ and a filled doner band below $E_f$. Arrows indicate how the electrode valence band shifts up with negative charging and down with positive charging. $E_f$ marks the highest filled energy level and is the Fermi energy of a thin slab several atoms thick representing the electrode surface. Reproduced from "How Computational Theorists Deal with Electrode Potential," Alfred B. Anderson 2025 J. Electrochem. Soc. 172 036,501 © 2025 Alfred B. Anderson. Published on behalf of The Electrochemical Society by IOP Publishing Limited

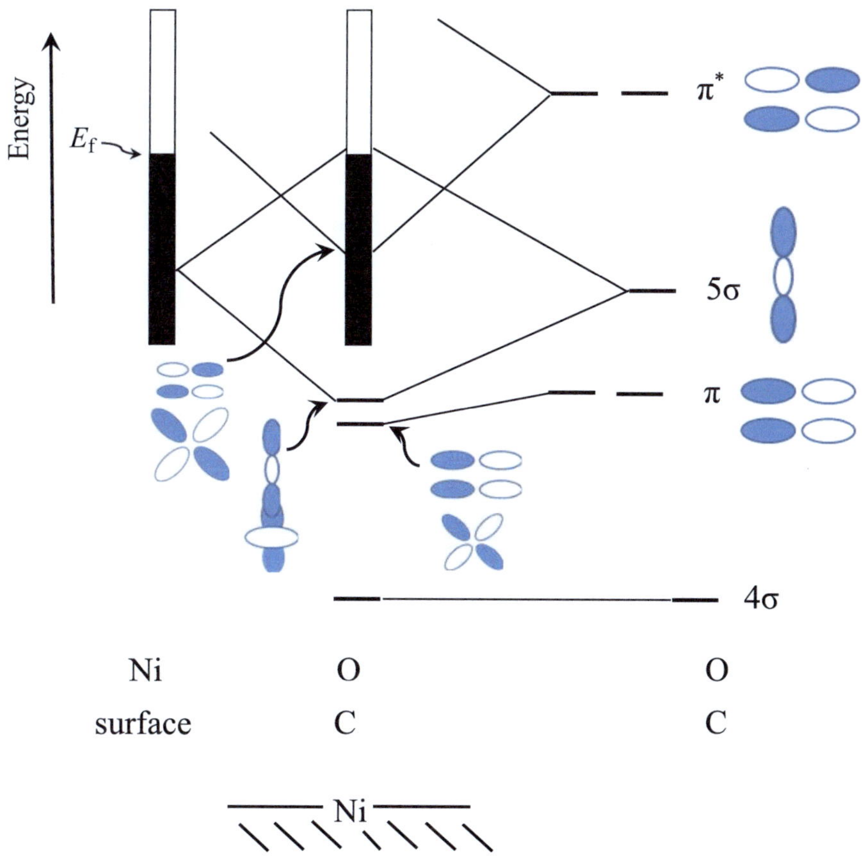


**Fig. 5.5** MO correlation diagram showing how CO valence MOs on the right mix with Ni valence band orbitals on the left when coordinated onefold atop adsorption site on a Ni atom in a Ni surface. The CO $\pi$, $5\sigma$, and $\pi^*$ levels and orbitals are shown mixing with band orbitals to produce occupied $\pi$ and $5\sigma$ bonding levels in the middle from overlap with Ni d orbitals. The antibonding CO $5\sigma$ + Ni d MO is unoccupied. The CO $\pi^*$ bonding interaction with the Ni d band is occupied and the antibonding counterpart is not

These two viewpoints were in contention and Linus Pauling was a proponent for localized orbitals. Charles A. Coulson, a proponent of MO theory, concluded from photoelectron measurements: "The single result of most significance is the complete vindication of the molecular orbital description of a molecule," see Turner [26] who quotes this from Ballard [27]. W. A. Goddard continued using the localized valence bond viewpoint. In simple cases local orbitals may be transformed to delocalized ones and vice versa.

## 5.11 Perturbation MO Theory

As changes in potential cause $E_f$ and the metal valence band to move up or down, the bonding orbitals and energies in Fig. 5.5 change. From Hoffmann's perturbation MO theory comes a rule that the upper orbital mixes with the lower in a bonding way, stabilizing it and the lower orbital mixes with the upper, destabilizing it. The essential results are in Fig. 5.6 where it is shown that the larger the energy difference, $\Delta$, the smaller the orbital mixing and energy shift.

For CO adsorbed on a nickel electrode surface, negative charging moves the electrode band (Fig. 5.4) up and this decreases the energy gaps between the CO $\pi^*$ orbital and all the valence band orbitals that have non-zero overlap integrals with it. The increased mixing with this CO antibonding orbital weakens the bond, so that it lengthens and the vibrational force constant and vibrational frequency both are decreased.

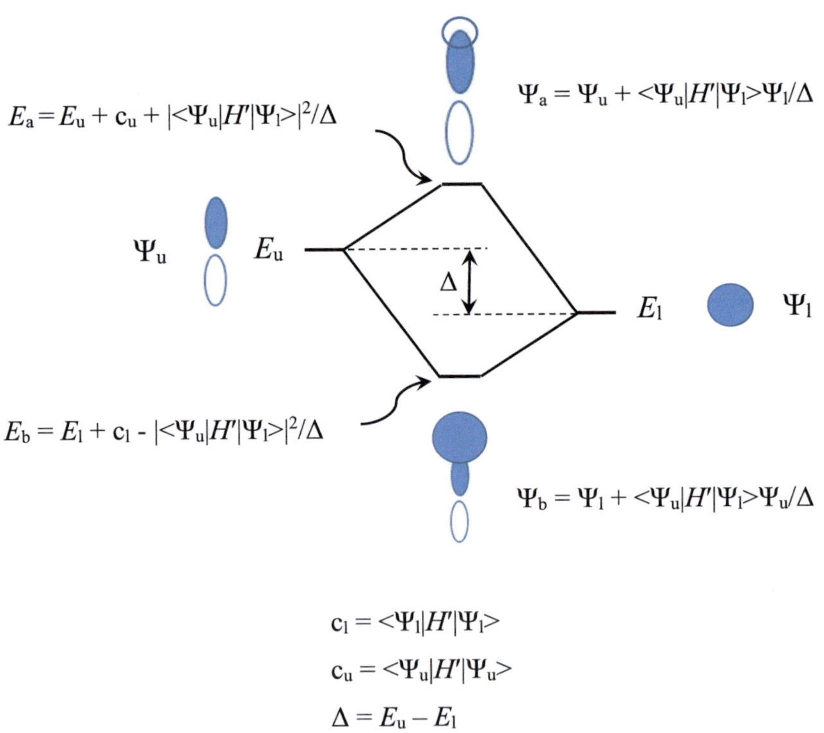


**Fig. 5.6** MO correlation diagram for a p orbital on an atom on the left with energy $E_u$ (upper) interacting with an s orbital on an atom on the right with energy $E_l$ (lower) to form σ and σ* orbitals. Qualitative perturbation formulas written on the diagram explain the rule of lower energy orbital stabilization by a bonding interaction and upper orbital destabilization by an antibonding interaction

## 5.12 Badger's Rule Relating Bond Length and Vibrational Frequency on Bond Length and Herschbach-Laurie Relationships

A relationship between bond length and harmonic force constants known as Badger's rule goes back to 1934 [28]. Herschbach and Laurie published systematic empirical relationships between cubic and quartic force constants and equilibrium bond lengths in 1961 [29]. Awareness of and understanding these relationships helps with understanding why molecular vibrational parameters trend across the periodic table and can depend on the molecule's environment.

## 5.13 Explaining Badger's and Hershbach-Laurie's Relationships by Partitioning Molecular Electron Distribution Functions into Atomic and Delocalization Components: A Charge Density Theory for the Dependence of Vibrational Frequency on Bond Length

A simple and conceptually useful theory for these relationships is based on partitioning molecular electron densities into rigid perfectly following atom-like ($\rho_a$, $\rho_b$) and delocalized non-perfectly-following ($\rho_d$) components that were defined by Anderson and Parr [30]. For a diatomic molecule ab, with $\rho_a$ centered on nucleus a at the origin and $\rho_b$ centered on nucleus b at position $R$ on the internuclear axis, the total molecular electron charge density function may be written as follows:

$$\rho_{ab}(\mathbf{r}, R) = \rho_a(\mathbf{r}) + \rho_b(|\mathbf{r} - R|) + \rho_d(\mathbf{r}, R) \tag{5.10}$$

the atom density functions are readily available from Hartree–Fock atomic orbitals in the literature and $\rho_d$ is approximated by a collection of rigid density distributions corresponding to bonds and lone pairs that do not overlap the nuclei. Then the divergence of the Hellmann–Feynman electrostatic force [31] on nucleus b, which is d$E$/ d$R$, where $E$ is the Born–Oppenheimer energy, equals the Laplacian of the energy with respect to movement of nucleus b and at the equilibrium internuclear distance, $R_e$ the harmonic vibrational force constant ($k_e$) falls out:

$$\begin{aligned}\left[\nabla^2_{R^2} E(R)\right]_{Re} &= 4\pi Z_b \rho_a(R_e) \\ &= \left[\mathrm{d}^2E/\mathrm{d}R^2 + 2/R\mathrm{d}E/\mathrm{d}R\right]_{Re} \\ &= k_e\end{aligned} \tag{5.11}$$

At the bottom of the vibrational potential energy, $E(R)$, curve the force term in the Laplacian, d$E$/d$R$, is zero and for the equilibrium internuclear distance, $R_e$, Eq. (5.11), which is why the equation reduces to a formula for $k_e$.

Taking derivatives of Eq. (5.11) and setting d$E$/d$R$ to zero each time gives formulas for cubic ($l_e$), quadratic ($m_e$), and higher force constants in terms of lower order constants and derivatives of $\rho_a(R)$.

For example, $l_e$ is given by the formula

$$l_e = 4\pi Z_b (d\rho_a/dR)_{Re} - 2k_e/R_e \tag{5.12}$$

Equation 5.12 in Chap. 3 shows how the force constants appear in the Taylor series expansion of the Born–Oppenheimer bond vibrational potential energy. Equation (5.11) is a classical electrostatic Poisson equation and it says the bond force constant is proportional to the atom nuclear charge multiplying the value of the rigid electron charge distribution centered on the other atom. Atomic charge density functions tail off with distance from the nucleus approximately exponentially and, within the model, this accounts for the Badger and Hershbach-Laurie relationships and also fits the behaviors of groupings of diatomic molecules across the periodic table [32]. As a rule, using the density of the heavier or more electronegative atom for $\rho_a$ in these equations gave more accurate predictions. Strong relationships were established between equilibrium bond lengths and the log of force constants for diatomic atoms using pairs of groups in the periodic table such as groups 4b and 6b, in this case holding up over a 1.5 Å range in equilibrium bond lengths, and for ground and excited state bond lengths and the log of force constants for CO [33]. The latter relationship implied that the oxygen valence charge density function behaved effectively independently from the electronic state CO was in and $k_e$ depended primarily on the CO bond length and, further, that $\rho_a(R)$ tailed off exponentially from the O nucleus. The relationship extended to bond stretching force constants for different electronic states of $CO_2$.

This model provided a basis for the bond force constants for stationary electronic states being a well-defined function of internuclear distance. Anderson also demonstrated relationships spanning pairs of columns of the periodic table of the elements derived from assuming $\rho_a$ in Eq. (5.11) is proportional to the product $Z_aZ_b$. These findings have been returned to recently by the application of machine learning [34]. Referring to the Anderson and Parr models, Mahmoud et al. noted "Most significantly, their work pointed toward a possible direct connection between a diatomic molecule's spectroscopic properties and its individual atoms' atomic properties and positions in the periodic table" [34]. After their processing by machine learning using Gaussian process regression, Mahmoud et al. conclude their paper with the following interesting statement: "Finally, it is worth mentioning that we are approaching a period in which machine learning techniques are as accurate as ab initio quantum chemistry methods for calculating spectroscopic constants of diatomics with almost no computational effort" [34]. Parameterized equations continue to be developed for diatomic bond properties, for example the work of Crittenden and references therein [35]. Bond lengths, strengths, harmonic force constants, and anharmonicities are all interconnected and electrochemical measurements of the dependence of vibrational frequencies on electrode potential tells a lot about a molecule at the electrochemical interface.

To digress briefly, R. T. Sanderson was successful at discovering predictive relationships not requiring quantum calculations, for example between electronegativity and bond energy [36]. Using available atomic density distribution functions, applications of Eq. (5.11) are in this spirit. It is noteworthy that it has not been explained quantum mechanically how the charge density tail of atom a is apparently intact when the nucleus of atom b is embedded in it, or even if this as the right way to think about the phenomenon. For more discussion, see Anderson [37]. Theorists can be slow initially at accepting such findings as meaningful. Sanderson addressed this point by saying "I suspect they [theorists] tend to judge such work by their own strange standards: (1) if the idea is simple enough to be understood, it cannot possibly be correct, and (2) if it consistently produces quantitatively accurate results, this must be entirely fortuitous because obviously it neglects so many important essentials that there is no chance of accuracy except by the purest of luck." [38] With the advent of machine learning, what is the future role of theory? Theory will provide understanding of structures, reactivity etc. in terms of the language of chemistry. If the desire to understand fades in time, chemistry as a discipline will involve laboratory and industrial synthesis, more in the spirit of chemical engineering.

In summary of the above, there is a strong relation between a bond's length and its force constant. The bond ground state classical vibrational frequency ($\omega_c$) is given at the harmonic level by

$$\omega_c = (k_e/\mu)^{1/2} \tag{5.13}$$

where $\mu$ is the reduced mass of the molecule. When CO is adsorbed on nickel, back-donation mixing of occupied Ni band orbitals with the empty CO $\pi^*$ orbitals causes the CO bond to weaken and lengthen, which decreases $\omega_c$. When the Ni is an electrode, negative charging should increase back donation to the $\pi^*$ orbitals and provide explanation for the electrochemical vibrational tuning rate. The 5σ donation may also change and have an effect on $\omega$.

Classically, the frequency in Eq. (5.13) is the vibrational frequency of a harmonic spring with force constant $k_e$. In the quantum case $\omega_e$ is not a vibrational frequency but instead is the difference between the energy of the first vibrational state, $E_1$, and the zero-point energy, $E_0$:

$$\omega_e = (E_1 - E_0)2\pi/h \tag{5.14}$$

Nevertheless, $\omega_e$ is still called a frequency. Born–Oppenheimer energies of chemical bonds are anharmonic, that is the higher order force constants are not zero, and for most bonds this makes the energy gap $E_1$-$E_0$ less than it would be when the harmonic level of approximation is assumed. Anharmonic contributions generally also decrease the classical harmonic oscillator value $\omega_c$ given by Eq. (5.13). Most of the lowering is accounted for as a function of the cubic force constant. It is further noted that the Taylor series of Born–Oppenheimer energy about $R_e$ has a radius of convergence equal to the length of $R_e$ because the potential is singular because of the coulomb repulsion between the nuclei as the internuclear distance approaches zero. This means that as the internuclear distance approaches $2R_e$ the energy expansion series using force constants produces a deteriorating fit. Nevertheless, the spectroscopic description of molecules is in terms of harmonic force constants sometimes augmented by the cubic or quartic level contributions to the Born–Oppenheimer energies is taken to be satisfactory in most cases.

Theories for the Stark tuning rates of free molecules include anharmonicity in the potential energy. Later discussion shows anharmonicity being incorporated in formulas derived to explain electrochemical vibrational tuning rates for adsorbed molecules in model electric fields. However, it may be pointed out, and will be discussed later, that quantum calculations of harmonic force constants for adsorbed molecules, whether by using atomic densities, semiempirical band shifting, ab initio Hamiltonians, or density functional theory, can be used to predict EVTR without including explicit electric fields.

Extended Hückel theory is not accurate enough for calculating the EVTR for CO by the band shift method because it does not include the strong electrostatic repulsive behavior resulting from atom overlap at short internuclear distances. Orbitals from EH calculations may be used to calculate electric fields but, because there is no actual charging of the surface, they will not have required potential dependence so that they cannot be used in the perturbation theory for gas phase molecules discussed by Bishop. Choosing an electric field model will be complicated by solvent molecules which also contribute to the field at the electrode surface making it non-uniform whatever the choice of double layer field model. The MO bonding and electronic structure picture presented above gives perspective that is applicable to understanding EVTR. To advance in this spirit, can an easily applied semiempirical theory be developed which gives EVTR without recourse to electric fields? The next section shows how this was done by combining aspects of EH and the atom charge density model for vibrational force constants.

## 5.14 ASED-MO, a Semiempirical Theory for Structures and Bond Lengths that, Within Gurney's Electrode Valence Band Model, Predicts EVTR

While the metal band can be shifted by changing the AO energy parameters, giving a qualitative account for the electrode potential effect in EH theory, might a simple computational approach allowing calculations of structures and force constants be found? A step in this direction begins with the electronic charge partitioning in Eq. (5.10). Using it, integrating the Hellmann–Feynman force on nucleus $b$ gives $E(R)$, the Born–Oppenheimer vibrational potential energy function for $a$–$b$:

$$E(R) = Z_b\left[Z_a R^{-1} - \int \rho_a(\mathbf{r})|\mathbf{r} - R|^{-1}\mathbf{dr}\right] - Z_b \int_\infty^R \int \rho_d(\mathbf{r}, R')d/dR'|\mathbf{r} - R'|^{-1}d\mathbf{r}/dR' \quad (5.15)$$

The terms in the square brackets comprise the electrostatic interaction energy between nucleus b, charge $Z_b$ and nucleus a, charge $Z_a$ and the spherical electron charge density cloud $\rho_a(\mathbf{r})$. This is repulsive because the spherical electron density can only partially shield the nuclear charge at the center, which is a well-known property if inverse $R$ potentials. The double integral term gives an attractive energy for bound molecules that comes from the integrating the electrostatic force on nucleus b caused by the delocalization electron charge density cloud which changes shape as the atoms superimpose. The energy of nucleus b interacting with $\rho_b$ is independent of $R$, and set to zero. The two components add to give the Born–Oppenheimer molecular bonding energy in Eq. (5.16), which is shown schematically in Fig. 5.7. Equation (5.16) is exact if the electron densities in Eq. (5.10) are exact

$$E(R) = E_r(R) + E_d(R) \quad (5.16)$$

Figure 5.8 shows how $E_r(R)$ increases as $\rho_a(\mathbf{r})$ and $\rho_b(|\mathbf{r}–R|)$ are superimposed as the internuclear distance is decreased. For heteronuclear molecules, the energy components $E_r(R) + E_d(R)$ each change when atoms a and b are exchanged, but the sum $E(R)$ does not. In applications of the model to calculating bond stretching force constants and bond strengths discussed below, accuracy was found to be greater when $E_r(R)$ was calculated using $Z_b$ for the less electronegative atom and $\rho_a(\mathbf{r})$ for the more

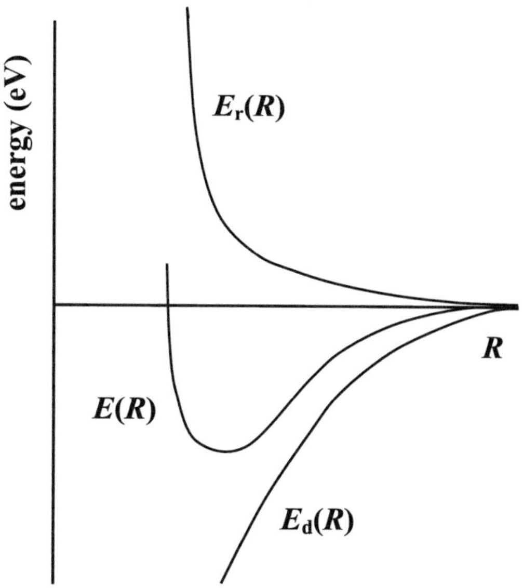


**Fig. 5.7** This schematic energy versus internuclear distance graph shows how the increasing repulsive component $E_r(R)$ and decreasing attractive component, $E_d(R)$ added to give a total Morse potential-shaped energy $E(R)$ as in Eq. (5.16)

**Fig. 5.8** This diagram shows circular cross sections of rigid charge density distributions for atoms a with its nucleus at $R = 0$ and b far away at first and then coming to overlap atom a as the internuclear distance $R$ decreases. Also drawn as a function of R is a graph showing how the repulsive electrostatic interaction of nucleus a with atom b increases as R decreases

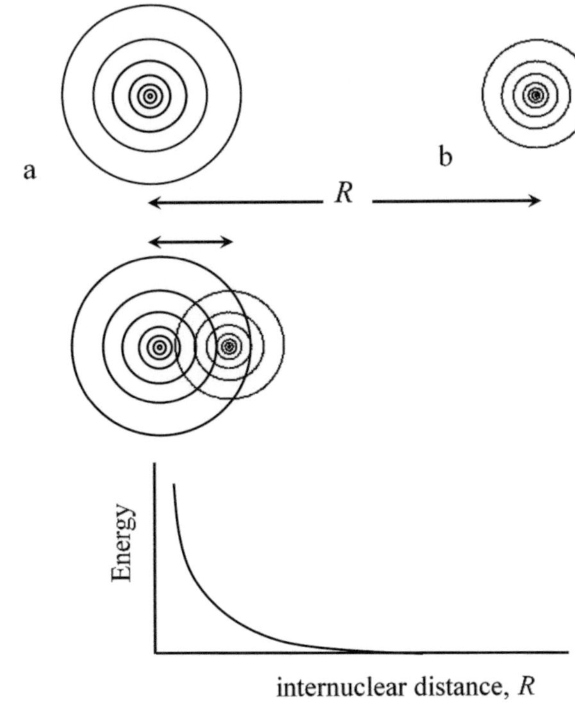


electronegative atom. The reason is not understood but more electronegative atoms in diatomic bonds generally gain charge making them larger and this is probably a factor.

To determine $\rho_d(\mathbf{r},R)$ it is necessary to find $\rho_a(\mathbf{r})$, $\rho_b(|\mathbf{r}-R|)$, and $\rho_{ab}(\mathbf{r},R)$ quantum mechanically and then evaluate

$$\rho_d(\mathbf{r}, R) = \rho_{ab}(\mathbf{r}, R) - \rho_a(\mathbf{r}) - \rho_b(|\mathbf{r} - R|) \tag{5.17}$$

When graphed $\rho_d(\mathbf{r},R)$ is a density difference map, typically with depletion of electron density around the nuclei and density buildup in the internuclear region of the chemical bond. In homonuclear molecules $\rho_d(\mathbf{r},R)$ is symmetric and in heteronuclear molecules there is electron density transfer to the more electronegative atom. In ionic molecules the anion gains density at the expense of the cation. Figure 5.9 shows $\rho_d(\mathbf{r},R)$ for a heteronuclear diatomic with ionic character. It is modeled after pictures based on Hartree–Fock calculations for LiF [39].

The attractive energy $E_d(R)$ depends on the delocalization of electrons from atoms into the space between the nuclei where they provide attractive bonding forces on the nuclei, themselves shielded by electrons in core orbitals. It was noted that MO energies calculated with the extended Hückel Hamiltonian are also delocalization energies associated with forming chemical bonds. Simply substituting the MO electronic energies for $E_d(R)$ in Eq. (5.16) was successful [40]. The procedure is not

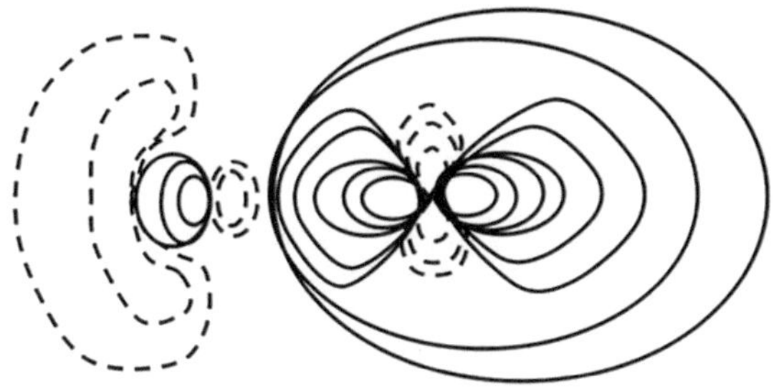

**Fig. 5.9** Simplification of the predicted density difference map for LiF of Bader, Henneker, and Cade [39]. Dashed lines indicate charge density depletion and solid lines indicate charge density increase. Charge contours for $F^{\delta-}$, on the right, show increased electron density compared to the atom and those on the left show depletion of the density on $Li^{\delta+}$

exact but the initial calculations gave reasonably accurate bond lengths in molecules. Predictions were improved by adding a factor in the Hückel Hamiltonian [41]. As in EH, diagonal Hamiltonian matrix elements are functions of the VSIP:

$$H_{\mathrm{ii}}^{\mathrm{a}} = -IP_{\mathrm{i}}^{\mathrm{a}} \tag{5.18}$$

Off-diagonal are slightly different from EH by having an exponential damping factor that was chosen to give optimal results for several molecules:

$$H_{\mathrm{ij}}^{\mathrm{ab}}(R) = (K/2)\left(H_{\mathrm{ii}}^{\mathrm{a}} + H_{\mathrm{jj}}^{\mathrm{b}}\right)S_{\mathrm{ij}}^{\mathrm{ab}}(R)\mathrm{e}^{-0.13R} \tag{5.19}$$

This procedure may be summed up as one of atom superposition and electron delocalization and with the electron delocalization energy being approximated by an MO energy from the above Hamiltonian. Hence it was named the ASED-MO theory. Overcoming limitations of extended Hückel calculations, with the inclusion of the $E_{\mathrm{r}}(R)$ energy component, ASED-MO calculations yield bond lengths, bond strengths and bond stretching force constants to useful accuracy for many problems. The theory uses valence Slater-type orbitals and the exponents and experimental VSIP are changed small amounts so that (i) the calculated bond lengths in diatomic molecules are close to experiment and (ii) the Mulliken charges on the atoms close to those predicted by differences in Pauling electronegativities for the atoms [1]. Following this mode of parameter selection, ASED-MO calculations of chalcogen atoms on small Ni clusters, employed as models for surfaces, showed good agreement with measured atom heights and the surface photoelectron spectrum for adsorbed O [42]. Bond strengths, lengths, and force constants calculated for first row transition metal diatomic oxides and sulfides from Sc through Cu were close to measured values [43, 44] Once parameters are determined, MO calculations on polyatomic systems proceed with summing the two-body repulsion energies atom pairs to give $E_{\mathrm{r}}$. Applications of the ASED-MO band shift approach to modeling EVTR and other electrode surface phenomena have been reviewed [45].

## 5.15 EVTR Calculations Using ASED-MO

The first study of EVTR was to provide an explanation for previously mentioned observations of R. Kötz and E. Yeager. This combined in situ Raman and ASED-MO study of $CN^-$ on a silver electrode [46]. Agreement between the theory and measurement was rough. In this work calculated values for bond force constants, not frequencies, were compared with experiment. For the $CN^-$ bond, the measured rate of force constant increase with potential was 0.38 mdyne/ÅV and the calculated value was 0.42 mdyne/ÅV. For the Ag-C bond force constant, the respective measured and calculated values were 0.15 mdyne/ÅV and 0.22 mdyne/ÅV. Figure 5.10 shows the electronic structure model and Fig. 5.11 compares calculated and experimental force constants as functions of electrode potential, labeled accidentally as eV rather than V in both figures. The errors in predicted frequencies are half of the errors in predicted force constants shown in Fig. 5.11. This is because, according to Eq. (5.13), $\omega_e$ is proportional to $k_e{}^{1/2}$. Expressed in terms of the measured quantity, $\omega_e$, the fit would appear better but the point is the ASED-MO theory gave values for the CN and AgC frequencies and approximate trends with relative potential. Criticisms can be made about the small four atom Ag cluster models, the omission of $CN^-$ adsorption at bridging sites, and the choice alignment of calculated force constants to the potential scale.

An ASED-MO study of CO modeling low coverage on Pt(111) predicted a ~65 $cm^{-1}$/V CO vibrational tuning rate [47]. Results of the calculation are in Fig. 5.12. The calculated EVTR for the PtC bond was small and negative. An experimental work at the time pegged the rate at ~30 $cm^{-1}$/V in acid electrolyte for the CO bond [48]. Coincidentally, 2/5 of this rate and 2/5 of the rate predicted by Holloway and Nörskov are both 30 $cm^{-1}$/V (see below). The experimental work mentioned above was probably for high CO coverage while the theories modeled low coverage. Later experimental investigations indeed show the EVTR increases with decreasing CO coverage [49, 50]. In the first of these the EVTR is 32 $cm^{-1}$/V at high, ~2/3 ML coverage, agreeing with the earlier measurements. For 0.45 ML and 0.43 ML coverages the respective rates for the two studies are 38 $cm^{-1}$/ V and 39 $cm^{-1}$/ V. For low coverages of about 0.1 ML the Chang and Weaver rate converges to 42 $cm^{-1}$/V while for Chen et al. the rate converges to 95 $cm^{-1}$/V. The disagreement may have been caused by CO adsorbing in relatively high coverage islands on the (111) electrode surface in the former case, which would favor a lower rate while in the latter study a polycrystalline film was used for the electrode and adsorption may have been on isolated low-coordinate sites. Due to the isolation the adsorbed CO molecules would not interact with each other. In any event, the measured EVTR increased for low coverages, becoming closer to the calculated values.

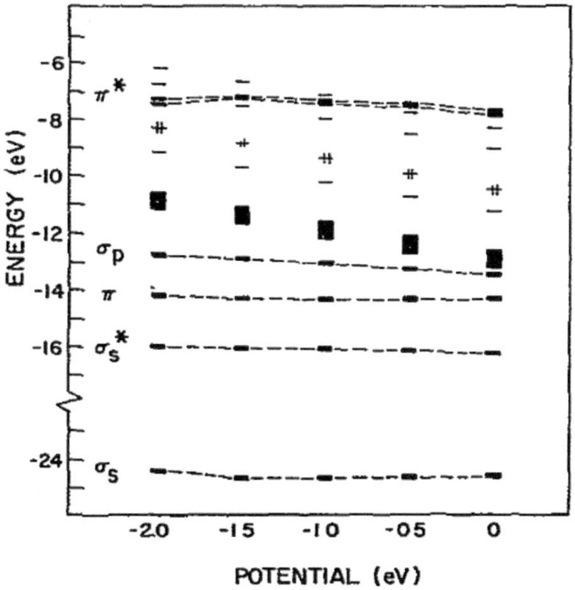


**Fig. 5.10** This shows the MO energy levels of $CN^-$ are shifted when CN bonds to a silver atom in an electrode surface. The occupied $CN^-$ sigma p energy level is increasingly stabilized as the silver valence band is stabilized, which models an increasing applied potential. Since this orbital is antibonding, more positive potentials delocalize it into the metal band and strengthen the CN and Ag-C bonds. Reprinted from Chemical Physics Letters, Vol 82, A. B. Anderson, R. Kötz, and E. Yeager, Theory for C-N- and Ag-C Vibrational Frequency Dependence on Potential: Cyanide on a Silver Electrode, 130–134, Copyright 1981, with permission from Elsevier

## 5.16 Alternative Approach of Holloway and Nörskov for Explaining EVTR

Holloway and Nörskov proposed a model where the density of states of adsorbed CO was assigned a Lorenztian distribution whose peak shifted linearly with potential change. The shift ascribed to a changing potential-dependent electric field, but $E_f$ was fixed. Though $E_f$ actually does change with increased potential, this model nevertheless gave trends in the electrochemical vibrational tuning rate because the CO $\pi^*$ moved further from $E_f$ as the potential increased within the model [51]. Here we see two approaches, both using bonding theory and one focusing on the effects of changing the surface $E_f$ and the other assuming a fixed $E_f$ and focusing on effects of changeable electric fields introduced as perturbations.

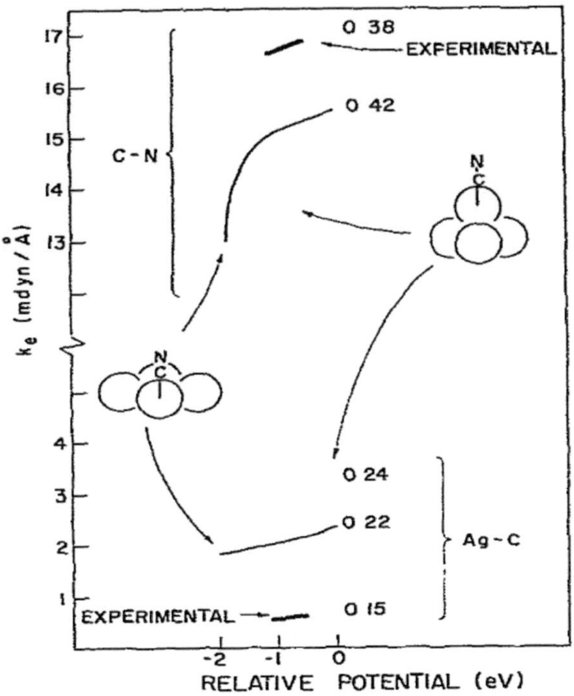


**Fig. 5.11** The magnitude of a bond force constant is a measure bond strength. This shows how $CN^-$ and Ag-C force constants increase as the electrode potential is increased and their magnitude and their rates of change agree qualitatively with measured values derived from measured vibrational frequencies over a potential range. Reprinted from Chemical Physics Letters, 82, A. B. Anderson, R. Kötz, and E. Yeager, Theory for C-N- and Ag-C Vibrational Frequency Dependence on Potential: Cyanide on a Silver Electrode, 130–134, Copyright 1981, with permission from Elsevier

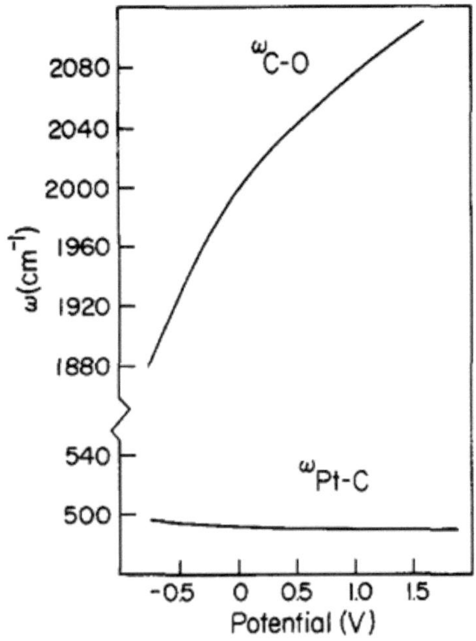


**Fig. 5.12** The graph shows how the CO vibrational energy when adsorbed on a Pt(111) surface model electrode with 19 atoms in the top layer and 12 atoms in a second layer, as calculated with semiempirical ASED-MO theory, increases with potential. Reprinted with permission from S. P. Mehandru, and A. B. Anderson, "Potential-induced variations in properties for carbon monoxide adsorbed on a platinum electrode," J. Phys. Chem. 93, 2044–2047, 1989. Copyright 1989 American Chemical Society

## 5.17 Applying ASED-MO to Demonstrate that Adsorption Sites of CO on Pd and Pt in Vacuum are Functions of the Metal $E_f$ Values and How Shifting Them Makes Pd Adsorb CO Like Pt and Makes Pt Adsorb It More Like Pd

Another illustration of the significance of band position on the energy scale lies in the explanation of CO(ads) site preferences on Pd and Pt (111) and (100) surfaces based on ASED-MO calculations [52]. Following what had become a well-established procedure experimental VSIP, were changed to give charge transfer between atoms in diatomic PtO and PdO in agreement with ionicity predictions based on Pauling's electronegativities. The calculated atom charges were determined using the Mulliken partitioning of the MO- based electron density distribution function [53]. Two-layer thick clusters were used to model the metal surfaces. The sites and cluster models are in Fig. 5.13.

The experimental Pd 4d VSIP is 0.8 eV less than the Pd 5d value [54]. This puts the Pd valence d band and $E_f$ approximately 0.8 eV higher in energy, so that the band orbitals mix more strongly with the CO $\pi^*$ orbitals. This is expected to favor bonding to bridging sites on the metal surfaces and from this observation the concept of structure factor may be defined. The difference between Pd and Pt VSIP holds up after adjustments for iconicity, with a value of 1.0 eV. Thus calculations favored the threefold site on Pd(111) and the twofold bridging site on Pd(100) Calculated energies

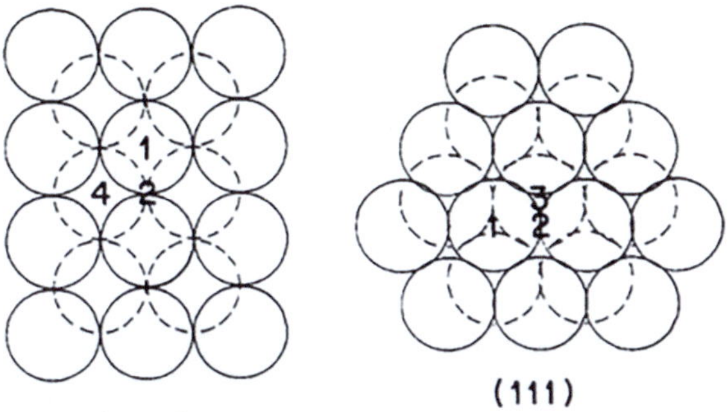


(100) (111)

**Fig. 5.13** On the left, a bi-layer cluster model for (100) surfaces with 12 atoms on top and six more in a second layer. On the right, a bi-layer cluster model for (111) surfaces with 12 atoms on top and seven more in a second layer. Nearest neighbor internuclear distances are constrained to the experimental bulk values, 2.74 Å for Pd and 2.77 Å for Pt. Unpaired electrons were assigned to the clusters, 14 for Pd and 18 for Pt in accordance with the ASED-MO procedure of occupying upper d band orbitals with at least one electron while lower ones are doubly occupied. Reprinted with permission from A. B. Anderson and Md. K. Awad, "Factors Determining CO Adsorption Sites on Pd and Pt (100) and (111) Surfaces: Theoretical study," J. Am. Chem. Soc., 107, 7854–7857, 1985. Copyright 1989 American Chemical Society

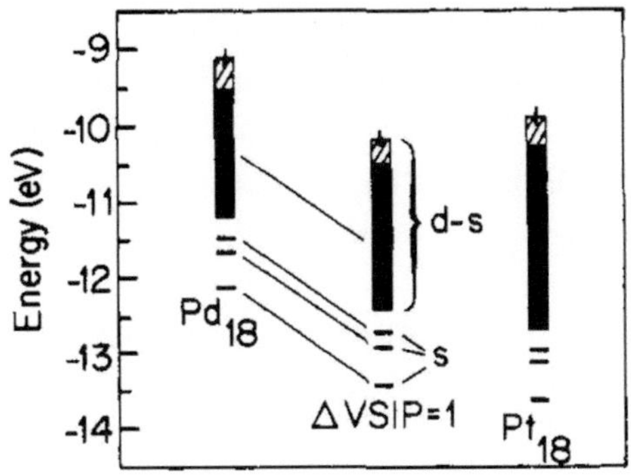


**Fig. 5.14** This shows how the Pd valence band of orbital energies when stabilized by 1 eV becomes very similar to that of the Pt band. Cross-hatching at the tops of the bands indicate ranges of energies where orbitals are singly occupied. Reprinted with permission from A. B. Anderson and Md. K. Awad, "Factors Determining CO Adsorption Sites on Pd and Pt (100) and (111) Surfaces: Theoretical study," J. Am. Chem. Soc., 107, 7854–7857, 1985. Copyright 1989 American Chemical Society

favored the onefold atop site on Pt(111) and on Pt(100) onefold atop and twofold bridging were equally stable. These theoretical predictions agreed with the findings reported in experimental literature. Most revealingly, when the Pd VSIP, adjusted for PdO iconicity, were increased by 1.0 eV, Pt structure predictions resulted and decreasing the adjusted Pt VSIP yielded Pd structures for adsorbed CO. Figure 5.14 shows how the 1.0 eV shift for Pd gives a valence d band position similar to that for Pt and Fig. 5.15 shows how the adsorption energies to 1- fold sites change with electrode potential.

It is concluded that the calculated difference between Pt and Pd vis-à-vis the adsorption of CO at the vacuum interface apparently depends on shifts in the metal d band position and Fermi energy without the need to invoke an influence of electric fields. As discussed above, such shifts also account at least qualitatively for the EVTR of CO on Pt(111). Furthermore, the Fermi energy is known from the work of Kötz, Neff, and Müller described earlier in this chapter to change proportionately with changes in the potential applied to the electrode. However, following the next section, models that use electric field are presented, leading up to the recent work involving vibrational chromophores tethered to electrodes and not bonded to them directly. Attempts are ongoing to sort out the relative contributions of potential-dependent inductive through-bond-coupling and electric field in systems such as these.

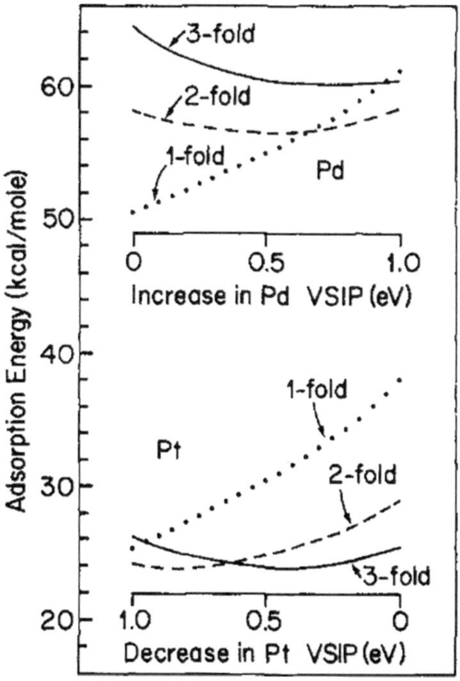


**Fig. 5.15** The upper part of the figure shows how as the Pd(111) surface valence band is stabilized, the adsorption bond energy of CO when on the onefold top site increases and surpasses bond energies for twofold and threefold bridging sites, giving the surface adsorption properties of P1(111). The lower part of the figure shows how as the Pt(111) surface valence band is destabilized CO adsorption energy at the onefold top site decreases until it becomes weaker than for the 2- and threefold sites when energy of the Pt valence band is similar to Pd. Reprinted with permission from A. B. Anderson and Md. K. Awad, "Factors Determining CO Adsorption Sites on Pd and Pt (100) and (111) Surfaces: Theoretical study," J. Am. Chem. Soc., 107, 7854–7857, 1985. Copyright 1989 American Chemical Society

## 5.18 Preview of First Principles Density Functional (DFT) Calculation of EVTR: CO on Pt(111)

The independent variable for EVTR measurements is the applied potential. If a Hamiltonian can be set up where the workfunction of a surface is changed by adding charge to it and this charge is compensated by a polarized ion double layer, then it becomes possible to calculate EVTR's as functions of applied electrochemical potential which is determined from the calculated $E_f$. A first principles code was developed for applying this idea to adsorption on electrodes [55]. The electric field is in principle a calculable but spatially complex function of position in the double layer while the measurable variable for presenting results of EVTR measurements is electrode potential. More will be said about the details of the first principles theory later. It was used to predict the EVTR in Table 5.1 for CO adsorbed on Pt(111) at coverages of 1/12, 1/6, ¼, ½, and 2/3 ML [56].

Predictions for higher coverage from about 0.66 to 0.4 ML correspond roughly to the experimental results of Chang and Weaver [49] for adsorption on Pt(111) and Chen et al. [50] for adsorption on polycrystalline Pt. The EVTR are highest for low coverage and the theoretical results become closer to the experimental results of Chen et al. at lower coverage, though at the lowest coverages the roughness of the polycrystalline film may contribute low coordinate sites with high EVTR. The low EVTR values of Chang and Weaver for the lowest coverages suggests CO islands of some average intermediate coverage may have formed. Further discussion is in the paper by Uddin [56]. Overall, this study brings up the uncertainties that often evolve when comparing measured surface properties in electrochemistry with predictions from computational theory.

As can be gathered from the work of Lambert presented in the next section and the work he references, attempts to interpret EVTR measurements of d$\upsilon$/d$U$ in terms of electric fields at the electrode surface and their dependence on the applied potential has given rise to a flurry of studies. Most older studies assume potential-dependent electric field models for their calculations. Implementing surface charging with double layer polarization and $E_f$ shifts as in the work outlined above came later. Lambert and others combined chemical bonding effects and electric field effects to extend the traditional viewpoint on Stark tuning into the electrochemical environment. However, the ASED-MO EVTR value for CO on Pt(111) at low coverage, 58 cm$^{-1}$/V, is close to the first principles value in Table 5.1, suggesting that for small adsorbed molecules the perturbation MO theory-based changes in bond strengths as functions of changes in Fermi energy of the surface dominate EVTR. Furthermore, the results from charge self-consistent first principles calculations for high coverage match well the experimental values. Neither method adds an external field perturbation to the Hamiltonian. Unfortunately, ASED-MO calculations have not been performed for high CO coverage to see if the predicted EVTR by

**Table 5.1** EVTR, d$\upsilon$/d$U$ (cm$^{-1}$/V), as functions of coverage, $\Theta$. Measured results by Chang and Weaver, first column; Chen et al. second column; self-consistent density-functional theory, third column

| Chang and Weaver [49] | | Chen et al. [50]. | | Theory [56] | |
|---|---|---|---|---|---|
| $\Theta$ | d$\upsilon$/d$U$ | $\Theta$ | d$\upsilon$/d$U$ | $\Theta$ | d$\upsilon$/d$U$ |
| ~0.6 | 32 | – | – | 0.66 | 32 |
| 0.45 | 38 | 0.43 | 39 | 0.50 | 43 |
| 0.35 | 40 | 0.30 | 62 | 0.25 | 62 |
| 0.20 | 42 | 0.16 | 82 | 0.17 | 68 |
| 0.12 | 42 | 0.13 | 95 | 0.08 | 70 |
| | | 0.09 | 95 | | |

this method decreases to ~32 $cm^{-1}$/V as found both in the measurements and first-principles predictions. For anyone interested in exploring the range of applicability of semiempirical methods, this would be a worthwhile study to make.

## 5.19 Electric Field Models of Bishop and Lambert

The initial emphasis on electric field in explaining EVTR stemmed from knowledge of the Stark effect in gas phase molecules, where vibrational frequencies were found to be changed by electric fields. So-called Stark tuning rates for adsorbed CO at the vacuum interface have drawn much attention and there have been attempts to connect the electrochemical phenomena to the vacuum interface observations. Bishop's perturbation theory provided formulas for frequency shifts that included cubic terms in the Born–Oppenheimer potential expansion and the quadrupole moment interaction with the electric field. For free CO, calculated Stark tuning rates were comparable to those from finite field quantum chemical calculations in the literature [57]. Lambert made multiple contributions to these efforts and has summarized much of the literature up to 1995 in a review [58]. He proposed a formula for the STR in free CO that uses measured spectroscopic properties of the CO molecule. He expressed the vibrational potential, $E'(\Delta R, F)$, where $\Delta R$ is the displacement of internuclear distance from its equilibrium value, to first order in a static electric field of strength, $F$:

$$E'(\Delta R, F) = E(\Delta R) - \gamma F p(\Delta R) \tag{5.20}$$

Here $E(\Delta R)$, has been modified by the interaction of its electric dipole moment to give a new equilibrium distance and $p(\Delta R)$ is the dipole as modified by the electric fields and $\gamma$ is a screening factor. He expanded $E(\Delta R)$ and $p(\Delta R)$ in Taylor series:

$$E(\Delta R) = a_{00} + a_{20}\Delta R_0^2 + a_{30}\Delta R^3 + \cdots \tag{5.21}$$

$$-p(\Delta R) = a_{01} + a_{11}\Delta R + a_{21}\Delta R^2 + \cdots \tag{5.22}$$

From these equations the following can be derived:

$$\mathrm{d}v/\mathrm{d}F = \gamma v_0(2a_{20}a_{21} - 3a_{30}a_{11})/\left(4a_{20}^2\right) \tag{5.23}$$

Here $\upsilon_0$ is the harmonic oscillator vibrational frequency. Using the known spectroscopic $a$ parameters Lambert calculated a vacuum STR of 42.8 ± 0.6 $cm^{-1}$/(V/Å). The ab initio calculations at the time yielded values scattered around this, ranging from 7 $cm^{-1}$/(V/Å) to 72 $cm^{-1}$/(V/Å). When using the experimental values for the cubic force constant, the second term in Eq. (5.23), dominates, meaning that according to this treatment the anharmonicity of the Born Oppenheimer vibrational potential dominates the STR.

Investigations were performed using the parameters in Eq. (5.23) to apply it to CO adsorbed on Pt by having the electric field vary with position. At the time ab initio work on adsorbed CO was for Cu and Pd surfaces but not Pt. In the ab initio studies, the applied fields and were found to affect the CO $\pi^*$ orbital occupations. The above references have further discussion of these findings.

Electroreflectance vibrational spectroscopy (EVS) was exploited at the vacuum interface to obtain STR for CO adsorbed on various surfaces and for Pt(111) at saturation coverage. Findings were 32 $cm^{-1}$/(V/Å), and at low coverage the value was $75 \pm 9$ $cm^{-1}$/(V/Å). These results seem *prima fascia* in agreement with results in Table 5.1. The question comes: how to properly compare these numbers with results of EVTR measurements and theory for electrode surfaces? As Lambert discusses, in attempting to do this it is necessary to make assumptions about the double layer field and the screening factor.

## 5.20 Electric Field Model of Mamatkulov and Filhol

The novel approach of Mamatkulov and Filhol begins with a model where the stretching frequency of adsorbed CO is taken to be a linear function of the surface charge, $q$ [59]. Electrochemical and mechanical contributions are postulated to the tuning rate, the latter being a linear function CO bond length. They express this with the following equation:

$$\omega(q, \Delta R) = \omega(0, R_0) + (\partial\omega/\partial q)_{R0} q + (\partial\omega/\partial R)_{q=0} \Delta R \tag{5.24}$$

The first differential term has charge dependence and is "electrochemical". It corresponds to the effect of electrode surface charge on the CO bond. The second differential term is "electromechanical" and stems from assuming that a positive electric field strengthens the bond by decreasing the internuclear distance, which will cause CO frequency to increase. This happens when the vibrational potential is anharmonic with the $l_e$ term in Eq. (3.12) of Chap. 3 having negative values, which is the case with the majority of bonds. Looking back at Eq. (5.11) it is clear that given the exponential increase in atomic electron charge density approaching the nucleus, if something serves to shorten the bond $\omega_e$ will increase. The formula was implemented by using the relationship for electric field between oppositely charged metal, plates $E = q/(\varepsilon_0 S)$ where S is a standard area. Using DFT, Mamatkulov and Filhol calculated energies of CO adsorbed to charged surfaces. To eliminate net charge in the translational unit cells a diffuse compensating charge cloud was added. They used the calculated electric field at the middle of the bond in the adsorbed CO molecules when parameterizing Eq. (5.24). For low coverage CO on Pt(111) they calculated approximate values for the two terms, obtaining $-6$ $cm^{-1}$ Å/V for the electrochemical one and 46 $cm^{-1}$ Å/V for the mechanical one. The sum, 40 $cm^{-1}$ Å/V equaled the value from total energy calculations with the same field model.

The total energy calculations for different fields also predicted that at high negative field values the adsorbed CO would shift from the top site to the threefold fcc site. The explanation given was that the calculated CO dipole was larger when in the threefold site and so with a strong negative electric field that site became favored. In contrast at the onefold site the calculated dipole was small and a positive field stabilized this site. The authors concluded that "Therefore, the CO stability is dominated by electrostatic interactions." This is in contrast to the view that band position controls the adsorbate–surface bonding energies as found earlier by Anderson and Awad. The fields and CO dipoles are derived quantities that may be evaluated using the variationally optimized wavefunctions but the energies are already at hand making these quantities superfluous.

Model fields introduced into the Hamiltonian are at best difficult to relate to the effects of changes in electrode potential. The majority of theoretical studies up until recently depended model electric fields in part because computational limitations precluded quantum mechanical treatment of the electrochemical interface with surface charging and electrolyte polarization. Entry into the expansive literature using fields is found in the above paper of Lambert and a more recent one [60]. In the latter, the authors conclude, based on using measured Stark shifting rates for CO on Pt(111) in vacuum and incorporating the field-dependent results in double layer field models for the electrochemical interface to predict EVTR, that a useful focus for future advancements will be more on chemical interactions involving potential-dependent bonding charge distributions than on local electric fields.

## 5.21 Role of Bond Polarizations, Hammett Parameter, Through Bond Coupling, and Electric Field on EVTR for Adsorbed and Distal Chromophores: Current Perspective

When molecules chemisorb on a metal, charge density distributions within them are changed, meaning the molecules are polarized, and surface charging changes the polarizations. Substituents, for example replacing a C-H bond with a C-F bond, can also polarize molecules. Relative polarizations can be anticipated using Pauling's electronegativity theory. The polarizations change charge distributions in a molecules bonds and will affect the electric fields around the molecule. As a prime example, polarizations can propagate through aromatic nitriles when dissolved in water and affect measurements of the solution pH. The Hammett parameter correlates with electron donating or withdrawing capability of the substituent, the value zero being assigned when a hydrogen atom is in the site where substitution is made. What is measured is the $pK_a$ of the chromophore. In terms of the value for the unsubstituted molecule, $K_0$, when H is in the site, and $K$, when the H is replaced by a substituent, the Hammett parameter, σ, is defined by

$$\sigma = \rho^{-1} \log(K/K_0) \quad (5.25)$$

Substituents polarize benzonitriles, affecting the CN bond strength and frequency. Measurements by Andrews and Boxer showed a rough correlation between Stark tuning rates and Hammett parameters for of nitriles immobilized in an organic molecular glass [61]. One may conclude that Stark vibrational tuning rates and Hammett parameters are both related to polarization changes caused by substituents.

There is growing interest in the possibility of measuring and calculating effects of changes in applied potential and changes in applied electric field on vibrations of bonds not directly attached to but near electrode surfaces. Many studies recently examined vibrational chromophores such as –CN tethered by molecular chains to an electrode with fields assigned based on surface charging and electrochemical double layer modeling. An investigation by Sakar et al. included 4-mercaptobenzonitrile bonded to a gold electrode through the S atom, with the CN not bonded to the surface [62]. When using an ionic liquid electrolyte, the measured CN vibrational frequency varied in a non-linear fashion over a 2 V range at a rate of about 4 $cm^{-1}$/V. This is 1/10th the rate compared to CN adsorbed on a silver electrode seen in the work of Kötz and Yeager that was discussed earlier.

Sarkar et al. measured CN vibrational frequencies for 20 para-substituted benzonitriles and when graphed as a function of their Hammett parameters a rough correlation appeared. They also set up a homogeneous electric field and calculated the frequencies for the benzonitriles, again obtaining a rough correlation with the Hammett parameters. They calculated the CN frequencies of benzonitrile and mercaptobenzonitrile as functions of introduced homogeneous electric fields and saw good linear correlations. The authors note that given the understanding that the CN frequency responds to polarizations of the substituted benzonitriles and because polarizations are induced by external fields or functional groups, they are justified in combining the data and express the field in terms of Hammett parameters. That is true for molecules enveloped by uniform electric fields, but for adsorbed molecules the induction due to bonding to the electrode must be considered and separation of the induction and field effects requires model assumptions. Thus the author's statement that "It is important to note that the interfacial electric field, and not the electrochemical potential, is the fundamentally important quantity in considering the polarization experienced by the tethered molecules" assumes minimal influence from inductive polarizations through the bonds resulting from bonding to the electrode surface. It is noted again that the field is a derived property that is a function of the electrode charge and its resulting potential. A significant question for chromophores tethered to an electrode surface is: At what tether length or chromophore distance from the surface does the induction effect approaches zero so that potential-dependent frequency shifts can be described in terms of the electric field alone?

In the study of Sarkar et al. the tether is short and while for the para-substituted benzonitriles the Hammett parameter values vary from −0.83 to 1.11 and the CN frequency varies by 22 $cm^{-1}$. As pointed out earlier, the EVTR for CN adsorbed on a silver electrode is 38 $cm^{-1}$/V. Comparing with the high rate for adsorbed CN is caused by induced CN charging and rehybridization, does the high rate for the delocalized $\pi$-bonded benzonitrile mean there is a significant inductive component? If so, the correlation between Hammett parameters and applied fields, while interesting, does not necessarily imply the field effect dominates over the inductive effect in the case of the benzonitriles.

Attempts are recently underway to separate inductive and field contributions to the Stark effect for self-assembled monolayers of mono-, bi- and terphenyl diisocyanide chains tethered to gold electrodes through a CN at one end [63]. In this case an aqueous electrolyte, 0.1 M $NaClO_4$ was used, and the tethered "free" CN, were found practically independent of the applied potentials while the CN bonded to the surfaces had high EVTR rates around 20 $cm^{-1}$/V. Similar results have been measured by others for gold electrodes [64]. From the near potential-independence of the frequencies of the free CN, Ge et al. concluded that in each case they lay outside of the Helmholtz layer. These authors also performed quantum chemical calculations with the diisocyanide chains bonded to small gold clusters and bathed in electric fields. They used a model in the form of Eq. (5.24) but with electric field replacing charge in the equation and, in linear extrapolation, calculated Stark tuning rates and found near cancellation of the small electromechanical and electric terms for the free CN. The bound CN values were larger and were dominated by the electromechanical term. The near zero value for the EVTR for the adsorbed mono- and bi-terphenyl diisocyanides is different from the 2.0 $cm^{-1}$/V reported by Sarkar et al. for the benzonitrile bonded through S to a gold electrode. This raises the question whether the CN group bonded to the surface in the diisocyanides absorbs most of the inductive effect.

A recent study of 4-mercaptobenzonitrile on gold employed a slab-band DFT surface model with dielectric continuum solvation, surface charging and electrolyte polarization [65]. These authors defined the electrode potential as $-E_F/e_0$ plus an electrostatic solution potential and the value zero was assigned for the potential of the uncharged slab. While compared with experimental measurements the calculated EVTR is too high, the main takeaway from the study is that the charge density and electric field maps show polarization of the whole molecule. At this time an understanding relative contributions of inductive and field effects on vibrational chromophores tethered to electrodes remains incompletely developed.

A review in 2021 gives perspective on field effects on chromophores in tethered molecules including discussion of much of the literature [66]. An ultimate practical goal for this research is to develop an understanding of the influence of electric fields on activating desired chemical reactions for electrochemical synthesis. As evident in the review and references therein, as well as the above discussion, this is a relatively new research topic for computational chemistry and increased understanding and applications lie ahead.

But it should be remembered that, as far as electrocatalysis on electrode surfaces is concerned, the electrode potential is the controlling variable for predicting and explaining reactions between adsorbed molecules. Theories that treat the electrochemical interface using electrode potential and Fermi energy can predict and explain electrocatalysis of adsorbed molecules and they also provide the charge distribution functions necessary for generating electric fields for ancillary analyses such as addressed in this section.

## References

1. Pauling L (1939) The nature of the chemical bond and the structure of molecules and crystals: an introduction to modern structural chemistry, Ithaca: Cornell University Press/London: Humphry Milford, Oxford University Press, 1939 and later editions
2. Simões JAM, Beauchamp JL (1990) Transition metal-hydrogen and metal-carbon bond strengths: the keys to catalysis. Chem Rev 90:629–688
3. Hammond GS (1955) A correlation of reaction rates. J Am Chem Soc 77(334–338):1955
4. Hansen WN, Wang CL, Humphries TW (1978) Electrode emersion and the double layer. J Electroanal Chem 90:137–141
5. Kötz ER, Neff H, Müller K (1986) A UPS, XPS, and workfunction study of emersed silver platinum and gold electrodes. J Electroanal Chem 215:331–344
6. Bagus PS, Pacchioni G (1991) Electric field effects on the surface-adsorbate interaction: cluster model studies. Electrochim Acta 36:1669–1675
7. Stark J (1917) Ermittlung des Spektrums eines einwertigen mehratomigen Molekülions, insbesondere des $H_2^+$-Ions. Ann Phys 52:221–254
8. Condon EU (1932) Production of infrared spectra with electric fields. Phys Rev 41:759–762
9. Bishop DM (1993) The vibrational stark effect. J Chem Phys 98:3179–3184
10. Koopmans T (1934) Über die Zuordnung von Wellenfunktionen und Eigenwerten zu den einzelnen Elektronen eines Atoms. Physica 1:104–113
11. Roothaan CC (1951) New developments in molecular orbital theory. Rev Mod Phys 23:69–89
12. Kötz R, Yeager E (1981) Potential dependence of vibrational frequencies of adsorbates on a silver electrode. J Electroanal Chem 123:335–344
13. Beden B, Lamy C, Bewick A, Kunimatsu K (1981) Electrosorption of methanol on a platinum electrode. ir spectroscopic evidence for adsorbed CO species. J Electroanal Chem 121:343–347
14. Kunimatsu K (1982) Potential dependence of infrared reflection adsorption spectrum of the adsorbed carbon monoxide produced by chemisorption of methanol on a smooth platinum electrode. J Electroanal Chem 140:205–210
15. Dewar M (1951) A review of π-complex theory. Bulletin de Société de Chimie Francein 18, c79
16. Chatt J, Duncanson LA (1953) Olefin co-ordination compounds. Part III. Infra-Red spectra and structure: attempted preparation of acetylene complexes. J Am Chem Soc, 2939–2947
17. Blyholder G (1964) Molecular orbital view of chemisorbed carbon monoxide. J Phys Chem 68:272–278
18. Hoffmann R (1963) An extended Hückel theory. I. Hydrocarbons. J Chem Phys 39:1397–1412
19. Robertson JC, Wilmsen CW (1972) Preferential orientation of CO adsorption on Ni as determined by extended Hückel calculations. J Vac Sci Technol 9:901–904
20. Grossmann A, Erley W, Ibach H (1995) Site occupation of CO adsorbed on Ni(100) at high CO pressures. Surf Sci 330:L646–L651
21. Anderson AB, Hoffmann R (1974) Molecular orbital studies of dissociative chemisorption of first row diatomic molecules and ethylene on (100) W and Ni surfaces. J Chem Phys 61:4545–4559
22. Döhnert D, Koutecky J, Schultze JW (1977) Applicability of the extended hückel method to a description of the electrosorption bond. J Electroanal Chem 82:81–92

23. Hoffmann R (1971) Interaction of orbitals through space and through bonds. Acc Chem Res 4:1–9
24. Turner DW (1970) Photoelectron spectroscopy. Ann Rev Phys Chem 21:107–128
25. Turner DW (1970) Molecular photoelectron spectroscopy: a handbook of He 584 a spectra. 1970 by John Wiley & Sons, 386 p
26. Turner D (1979) Molecular photoelectron spectroscopy. Nature 279:88
27. Ballard RE (1978) Photoelectron spectroscopy and molecular orbital theory. Adam Hilger, Bristol
28. Badger RM (1934) A relation between internuclear distances and bond force constants. J Chem Phys 2:128–131
29. Herschbach DR, Laurie VW (1961) Anharmonic potential constants and their dependence upon bond length. J Chem Phys 35:458–463
30. Anderson AB, Parr RG (1972) A poisson equation for vibrational potentials of diatomic molecules. Theoret. Chem. Acta 26:301–310
31. Feynman RP (1939) Forces in molecules. Phys Rev 56:340–343
32. Anderson AB, Parr RG (1971) Universal force constant relationships and a definition of atomic radius. Chem Phys Lett 10:293–296
33. Anderson AB (1972) On effective molecular electronic charge densities and vibrational potential energy functions. J Mol Spectroscopy, 411–424
34. Ibrahim MAE, Liu X, Pérez-Ríos J (2024) Spectroscopic constants from atomic properties: a machine learning approach. Digital Discovery, 3:34–50. https://doi.org/10.1039/d3dd00152k
35. Crittenden DL (2023) Simple, near-universal relationships between bond lengths, strengths, and anharmonicities. AIP Adv 13:115323. https://doi.org/10.1063/5.0175455
36. Sanderson RT (1983) Electronegativity and bond energy. J Am Chem Soc 105:2259–2261
37. Anderson AB (1994) Electron density distribution functions and the ASED-MO theory. Int J Quantum Chem 49:581–889
38. Sanderson RT (1986) Is the theoretical emperor really wearing any clothes? J Chem Ed 63:845–846
39. Bader RFW, Henneker WH, Cade PE (1967) Molecular charge distributions and chemical binding. J Chem Phys 46:3341–3363
40. Anderson AB, Hoffmann R (1974) Description of diatomic molecules using one electron configuration energies with two-body interactions. J Chem Phys 60:4271–4273
41. Anderson AB (1975) Derivation of the extended Hückel method with corrections: one electron molecular orbital theory for energy level and structure determinations. J Chem Phys 62:1187–1188
42. Anderson AB (1977) Determination of Chalcogen structures on nickel surfaces from orbital energies added to pairwise atomic repulsions. J Chem Phys 66:2173–2175
43. Anderson AB, Grimes RW, Hong SY (1987) Toward a better understanding of the atom superposition and electron delocalization molecular orbital theory and a systematic test: diatomic oxides of the first transition-metal series, bonding and trends. J Phys Chem 91:4245–4250
44. Anderson AB, Hong SY, Smialek JL (1987) Comparison of bonding in first transition-metal series: diatomic and bulk Sulfides and oxides. J Phys Chem 91:4250–4254
45. Anderson AB (1990) The influence on electrochemical potential on chemistry at electrode surfaces modeled by MO theory. J Electroanal Chem 280:37–48
46. Anderson AB, Kötz R, Yeager E (1981) Theory for $C\text{-}N^-$ and Ag-C vibrational frequency dependence on potential: cyanide on a silver electrode. Chem Phys Lett 82:130–134
47. Mehandru SP, Anderson AB (1989) Potential-induced variations in properties for CO adsorbed on a platinum electrode. J Phys Chem 93:2044–2047
48. Kunimatsu K, Seki H, Golden WG, Gordon JG II, Philpott MR (1985) Electrode/electrolyte interphase study using polarization modulated FTIR reflection-absorption spectroscopy. Surf Sci 158:596–608
49. Chang SC, Weaver MJ (1990) Coverage-dependent dipole coupling for carbon monoxide adsorbed at ordered Pt(111)-aqueous interfaces: structural and electrochemical implications. J Chem Phys 92:4582–4594

50. Chen YX, Heinen M, Jusys Z, Behm RJ (2007) Potential-induced $CO_{ad}$ island formation on a platinum thin-film electrode an the H-upd potential region. J Phys Chem C 111:435–435
51. Holloway S, Nörskov JK (1984) Changes in the vibrational frequencies of adsorbed molecules due to an applied electric field. J Electroanal Chem 161:193–198
52. Anderson AB, Awad MdK (1985) Factors determining CO adsorption sites on Pd and Pt (100) and (111) surfaces: theoretical study. J Am Chem Soc 107:7854–7857
53. Mulliken RS (1955) Electronic population analysis on LCAO-MO molecular wave functions. I. J Chem Phys 23:1833–1840
54. Lotz W (1970) Electron binding energies in free atoms. J Opt Soc Am 60:206–210
55. Jinnouchi R, Anderson AB (2008) Aqueous and surface redox potentials from self-consistently determined Gibbs energies. J Phys Chem C 112:8747–8750
56. Uddin J, Anderson AB (2013) Trends with coverage and PH in stark tuning rates on Pt(111). Electrochim Acta 108:398–403
57. Bauschlicher CW (1985) The effect of an external electric field on the vibrational frequency of CO. Chem Phys Lett 118:307–310
58. Lambert DK (1996) Vibrational stark effect of adsorbates at electrochemical interfaces. Electrochim Acta 41:623–630
59. Mamatkulov M, Filhol J-S (2011) An Ab Initio study of electrochemical versus electromechanical properties: the case of CO adsorbed on a Pt(111) surface. Phys Chem Chem Phys 13:7675–7684
60. Wang H, Tobin RG, Lambert DK (1994) Coadsorption of hydrogen and CO on Pt(335): structures and vibrational stark effect. J Phys Chem 101:4277–4287
61. Andrews SS, Boxer SG (2000) Vibrational stark effects of nitriles I. methods and experimental results. J Phys Chem A 104:11853–11863
62. Sarkar S, Patrow JG, Voegtle J, Pennathur AK, Dawlaty JM (2019) Electrodes as polarizing functional groups: correlation between Hammett parameters and electrochemical polarization. J Phys Chem C 123:4926–4937
63. Ge A, Videla PE, Lee GL, Rudshteyn B, Song J, Kubiak CP, Batista VS, Yian T (2017) Interfacial structure and electric field probed by *in situ* electrochemical vibrational stark effect spectroscopy and computational modeling. J Phys Chem C 121:18674–18682
64. Zhao K, Xiong H, Xiao Y, Su H, Wu D, Chang X, Lu Q, Xu B, Understanding the effect of specific adsorption on the vibrational Stark effect of adsorbates on an electrode surface *via* surface enhanced spectroscopy. Inorg Chem Frontiers. https://doi.org/10.1039/d3qi02122
65. Goldsmith ZK, Secor M, Hammes-Schiffer S (2020) Inhomogeneity of interfacial electric fields at vibrational probes on electrode surfaces. ACS Cent Sci 6:304–311
66. Bhattacharyya D, Videla PE, Cattaneo M, Batista VS, Lian T, Kubiak CP (2021) Vibrational stark shift spectroscopy of catalysts under the influence of electric fields at electrode-solution interfaces. Chem Sci 12:10131–10149

# Chapter 6
# Electronic Structure Picture of Anode, Cathode: Equilibrium with No Current and Conditions at Steady State with Current

**Abstract** Using the example of the hydrogen fuel cell in acid electrolyte, electronic structures are used to take into account the electrochemical potential of electrons on the cathode, anode, and the connecting wire. In the absence of ion flow in the electrolyte, the anode and cathode reactions with $H_2$ and $O_2$ are at equilibrium with the respective oxidation and reduction products. At a catalytic anode the electron chemical potential for the $H_2$ oxidation equilibrium establishes the thermodynamic work function and potential of the electrode surface. Similarly, at a catalytic cathode the electron chemical potential for the $O_2$ reduction equilibrium determines the potential of the electrode surface. Band bending is invoked to connect the surface and bulk electronic structures. When ion conduction in the electrolyte is turned on, electrons generated at the metal cathode enter the conduction band and protons enter the electrolyte. These combine with $O_2$ at the cathode where the electron is 1.2 eV more stable. Along the way the electron loses this energy by generating heat, light or electromagnetic energy. Polarizations at interfaces represented by band bending do not affect the electron energy relative to the cathode as it moves in the wire via the conduction band.

## 6.1 General Introduction

Gurney introduced band theory and adjustable metal surface electron chemical potential which has been the basis of the band shift models introduced so far. Often this chemical potential is called the Fermi energy ($E_\mathrm{f}$). For computations using slabs of metal atoms representing an electrode surface, this is $E_\mathrm{f}$ of the slab. But in the actual electrode the slab is interfaced with the bulk metal comprising the wire that conducts the generated current. Experimental and theoretical understanding of electrode reactions for the most part considers just the electrode with its reaction without taking into account the other electrode. Therefore, when focusing on a slab model of an electrode surface, semiempirical band shifting approaches to changing the potential do represent the changes in surface electron chemical potential of an electrode as do advanced first principle DFT methodologies described in later chapters. For a

 
A. Anderson, *Past and Current Models in Electrochemistry*,
https://doi.org/10.1007/978-3-031-97791-6_6

complete theory of an electrode, the slab must be interfaced with the semi-infinite bulk it sits on and the bulk metal has its own standard work function and electron chemical potential. This chapter characterizes the overall system with anode and cathode reactions with connecting wires and electrolytes. The hydrogen fuel cell is chosen as an example.

## 6.2 Electronic Structure of Atoms and Large Systems and Effects of Electric Fields

While it is clear experimentally that the work function of an electrode surface depends on surface conditions including adsorption and added charge, how is the electronic structure of the surface layer of atoms and of atom layers beneath affected? When there are adsorption bonds to the surface with no added charge, perturbation MO theory can account for shifts in work function. How to view work function changes accompanying charging the surface? Bigelow, using a model Hamiltonian, showed in 1979 that externally applied uniform electrostatic fields shift the orbital energy levels in a molecule uniformly and otherwise preserve its electronic structure [1]. Using the Janak theorem, in density functional theory the work function of an extended uniform surface such as Pt(111) is the negative of the highest occupied molecular orbital (HOMO) [2]. The electronic structure below the energy of the HOMO is preserved. The two models used for these conclusions are not exact but the conclusions give a frame for understanding surface electronic structures. Sparsely distributed adsorbates participating in electron transfer with the surface may give rise to local work functions that can be measured for vacuum interfaces by photoelectron spectroscopy [3]. On an electrode surface there may be local work functions and the structures and charge distributions that cause them may influence electrochemical reaction activity.

## 6.3 Electronic Structures of Electrodes at Steady State and During Current Flow Using the Hydrogen Fuel Cell as an Example

An electrode potential is set by the electron chemical potential which is determined by a reaction at equilibrium at the electrochemical interface. The anode equilibrium reaction

$$1/2H_2(g) \rightarrow H^+(aq) + e^- \tag{6.1}$$

at standard state conditions has a reaction Gibbs energy of zero when the electron energy is about −4.43 eV and the potential of the electrons is assigned as 0 V(SHE), which is the standard hydrogen electrode reference for electron transfer reactions.

For $O_2$ reduction at standard state conditions the cathode reaction

$$1/4O_2(g) \rightarrow H^+(aq) + e^- \rightarrow 1/2H_2O(l) \quad (6.2)$$

has reaction energy 1.229 eV when the electron energy is zero (on the vacuum scale) but when the electron has energy −1.229 eV, Eq. (6.2) is at equilibrium and the reversible potential 1.229 V(SHE). Electrode materials play no direct role in determining standard reversible potentials of electrode surfaces and when variations in work function across an electrode surface are small the equilibrium potential dominates electron transfer. For a steady state current flow to a counter electrode, anode and cathode surfaces will both need to have catalytic ability to allow reactions (6.1) and (6.2) to proceed.

Because the electron chemical potential of an electrode surface is generally not the same as the as the electron chemical potential of the bulk electrode material or the wire connecting the electrodes, the band bending concept is used to connect the electronic structures of the surface and bulk phases. This is illustrated in Fig. 6.1 which includes the Pt anode and cathode and a connecting wire taken to be platinum. The band bending in the illustration will extend only a few atom layers into the bulk because of shielding by the polarizable electron charge. It does not represent a barrier or energy loss for electrons passing through them. The electrode surfaces are at equilibrium but no current can flow because an electrolyte connecting the electrodes, which is necessary for proton transport, is not yet included. Standard potentials are shown and the −6.0 eV value for $E_f$ is from an average of several work function measurements for platinum [4–6] The workfunction is the energy of the vacuum level (0.0 eV) minus $E_f$. Any other conducting wire with a different workfunction could be used and at each metal junction and a band bending will ensue.

A schematic cross section of the physical system corresponding to Fig. 6.1 is shown in Fig. 6.2

The polarized electrolyte and charged surface are stable on emersion, as discussed in the previous chapter, maintaining the −4.4 eV electron chemical potential and 4.4 eV work function for the anode and the respective −5.6 eV and 5.6 eV values for the cathode. Measurements of the work local function in the shaded blue regions in the electrodes will give values between those of the surfaces of the electrodes and the and the surface of the wire outside the band bending region.

For electrons to flow from the anode to the cathode through the wire, charge must be balanced by having hydronium ions flow through the electrolyte. When this happens electrons move using states at the bottom of the conduction band as in Fig. 6.3. Here electrons travel through the wire and perform work such as generating heat, running an electromagnetic motor, or generating light before the exhausted electrons reach the cathode where they are extracted and react with $O_2$. The electron begins its travel at the anode and all along the wire the work function is 6.0 eV. The electron holds 1.2 eV potential energy relative to the cathode. It is difficult to represent the energy with customary band pictures, so the artificial vacuum level shift is invoked in Fig. 6.3. Figure 6.4 is a cross section view of the system showing

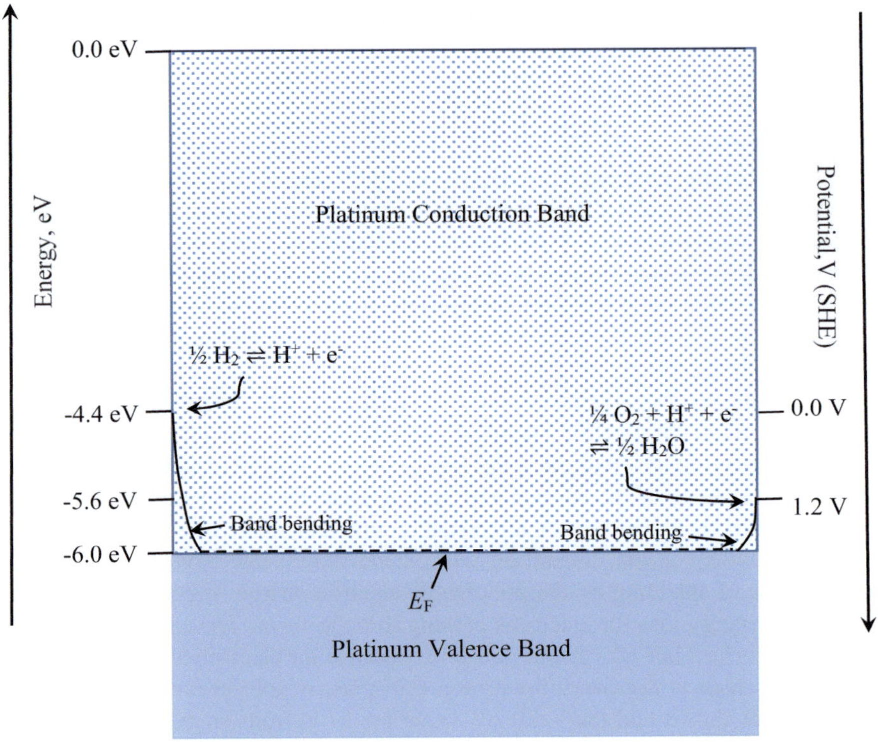


**Fig. 6.1.** Schematic energy diagram for a platinum anode and cathode connected by a platinum wire when there is no electrolyte bridge to conduct protons. Standard conditions are assumed. Hydrogen gas is oxidized in the left at the anode and is in equilibrium with the electrolyte and the surface of the electrode, setting the potential at 0.0 V. On the right oxygen reduction sets the potential at 1.2 eV. Platinum valence and conduction bands and the Fermi energy of the bulk platinum, $E_F$, which is at $-6.0$ eV are shown. Bending up of the conduction band to match the surface potentials at each electrode is indicated.

the electron pathway in the wire and the hydronium ion pathway in the electrolyte. Electron and proton currents flow as if on conveyor belts. As an electron and proton are formed at the anode another electron and proton are consumed at the cathode. For clarity, no resistance is shown in the wire. This is the workings of the hydrogen fuel cell.

Looking at Fig. 6.1 one might wonder if electrons at the anode can be ejected into the conduction band and travel through the metal in this band level. This corresponds to an excited state of the metal and there might not be a high enough density of states in this energy region for high conductivity. Furthermore, hot electrons have short mean free paths, in the nanometer region, and they would quickly de-excite and heat the metal, possibly even before exiting the band bending region [7]. Instead, adiabatic transfer puts the electrons in wave packets made up from empty states at the bottom of the conduction band and these packets are drawn to the cathode because

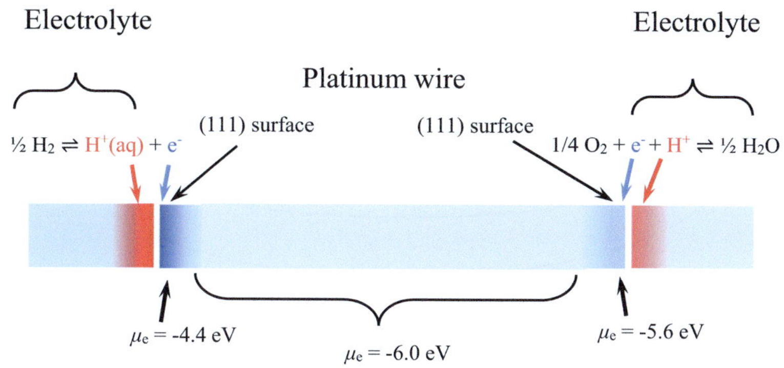


**Fig. 6.2.** Schematic cross section of two cells in acid electrolyte connected by a Pt wire. Shaded red regions indicate qualitative distribution of $H^+(aq)$ in the electrolyte near the electrode surfaces and shaded blue regions indicate electron polarization at the electrode surfaces. The electrolytes at the anode and cathode are not yet connected and no current flows.

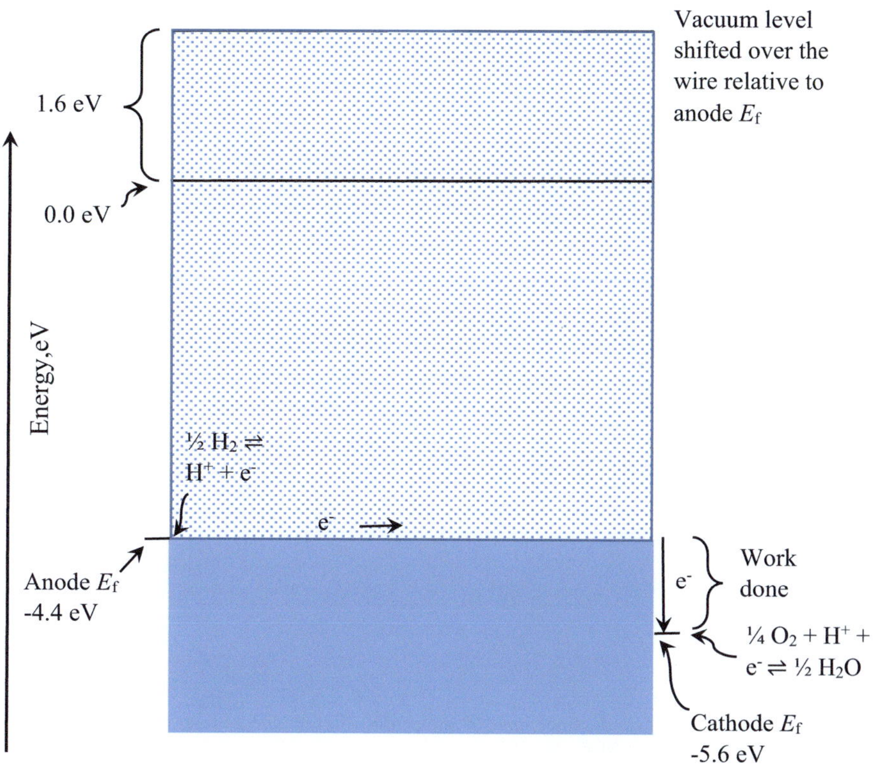


**Fig. 6.3.** As in Fig. 6.1 but with currents flowing, electrons through the wire and hydronium ions through the electrolyte. The electrons perform 1.2 eV of work during their passage from the anode at −4.4 eV to the cathode at −5.6 eV. Producing electromagnetic work, heat, or light also depletes the electron energy and when the electrons finally reach the cathode the energy is spent.

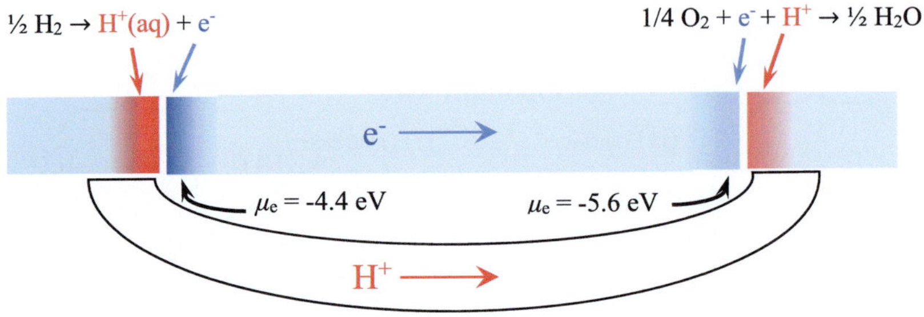


**Fig. 6.4** As in Fig. 6.2 including a schematic electrolyte bridge between the anode and cathode cells. Proton flow from the anode to the cathode through the bridge and electrons flow through the connecting wire.

of its lower free energy. To maintain the electronic structure viewpoint, the energy of the vacuum level over the wire is artificially shifted up by the energy the electrons possess relative to the cathode energy level, as in Fig. 6.3.

When n cells are connected in parallel the voltage measured is unchanged. When they are connected in series each cell contributes $-1.2$ eV to the reaction energy so that the voltage is 1.2(n) V. The potential of a cell can be adjusted to smaller values by a variable resistor. The potential may also be changed to accommodate non-standard conditions via the Nernst equation.

Consider for a moment the generation of $H_2$. Chapter 1 suggested the interaction between H atoms and the Pt surface may be strong. In fact, Chap. 10 introduces underpotential deposited (upd) H on Pt. When the potential is decreased through the double layer potential range H begins to adsorb at about 0.4 V(SHE) and the surface concentration builds as the potential approaches 0 V(SHE) when $H_2$(g) evolves. This is the equilibrium potential for the hydrogen anode. In contrast Cu is a relatively poor electrocatalyst for hydrogen evolution and, also introduced in Chap. 10, over potential deposited (opd) H begins to form at about $-0.2$ V(SHE) and only when the potential reaches $-0.5$ V(SHE) does rapid $H_2$(g) evolution commence. Since H atoms adsorb relatively weakly on copper, for H(ads) to be in equilibrium with $H_3O^+$ demands less stable electrons, meaning lower potentials. Consequently, a copper anode in a hydrogen fuel cell will have low activity at 0.0 V

## References

1. Bigelow RW (1979) The displacement of molecular orbitals due to interaction with charged species. J Chem Phys 70:3139–3140
2. Janak JF (1978) Proof that ∂E/∂ni = εi in density-functional theory. Phys Rev 18:7165–7168
3. Wandelt K (1997) The local workfunction: concept and implications. Appl Surf Sci 111:1–10
4. Derry GN, Zhang J-Z (1989) Workfunction of Pt(111). Phys Rev B 39:1940–1941
5. Bouwman R, Sachtler WMH (1970) Photoelectric determination of the work function of gold-platinum alloys. J Catal 19:127–140

6. Jinnouchi R, Anderson AB (2008) Electronic structure calculations of liquid-solid interfaces: combination of density functional theory and modified Poisson-Boltzmann theory. Phys Rev B 77:245417-1-18
7. Ievgen Nedrygailov II, Park JY (2016) The nature of hot electrons generated by exothermic catalytic reactions. Chem Phys Letters 645:5–14

# Chapter 7
# Double Layer Formation on Insulators

**Abstract** This chapter introduces concepts based in recent literature. Experimental measurements show that electrons at the valence band at the top of an insulator surfaces can be in equilibrium with near surface reactions in solution. The example is diamond. Two effects are demonstrated. For the first, oxygen and carbon dioxide in the ambient atmosphere oxidize the surface of diamond to form holes ($h^{+}_{dia}$), and bicarbonate anions. In this double-layer system the holes in the surface of the insulator are the charge carriers for p-type conduction laterally across the surface region. Second, the electrochemical potential of electrons in solution with oxygen present is determined by its four-electron reduction potential for forming $H_2O$ and has pH dependence determined from the Nernst equation. Depending on pH, electrons can transfer from or to the diamond surface until the chemical potential of electrons in the surface matches that in the solution so that equilibrium is established with the oxygen containing solution. The same mechanism can make band gap state occupations depend on pH, affecting optical properties of other insulators.

## 7.1 General Introduction

In the discussion of electrochemistry so far, the metallic electrode immersed in an electrolyte has been a part of a half cell and its potential is measured relative to a standard half-cell with a volt meter inserted in an external wire connecting the two cells. On making the connection, electrons pass and when steady state current has been achieved voltage is read. The potential of the electrode relative to a standard half-cell is a function of the concentrations of reacting species at its surface and is given by the Nernst equation. A double-layer is formed and its robustness was established in the emersion experiments of Hansen et al. [1] and Kötz et al [2]. The double-layer forms in the absence of a connection to another half-cell. As discussed in Chapter 6, the electrons can go from the electrolyte to a surface or from the surface to the electrolyte depending on the electrolyte composition.

When an insulating material is placed in water with dissolved ambient atmospheric gasses, there may be conditions where the surface becomes charged because

 
A. Anderson, *Past and Current Models in Electrochemistry*,
https://doi.org/10.1007/978-3-031-97791-6_7

of electrons transferring from the valence band and reacting with components of the solution, forming anions. In this chapter it is shown that under this circumstance the insulator may become a surface p-type conductor. It is also shown how the pH dependent electron chemical potential of electrons in solution may change the occupation of the surface states participating in the double-layer. By the same mechanism, changes in pH can also affect occupation of states in the pH-dependent potential range that might be extant in the gap between the valence and conduction band. Optical spectroscopic properties of the insulator will then become dependent on pH. This is a relatively new and possibly unfamiliar domain for some electrochemists and understanding and applications for these conducting insulator surfaces are evolving.

## 7.2 Inducing Conductivity in Semiconductor and Insulator Surfaces

When the surface of an insulator is interfaced with a material of sufficiently high electron affinity, electrons can transfer to form a double layer with holes ($h^+$) in the surface region of the insulator with the charge compensating donated electrons residing in a thin layer of accepting material. An insulator may have orbitals called surface states localized in the surface region with energies at the top of the filled valence band. As noted for semiconductors by Bardeen [3], a double layer may form using surface states of the semiconductor as electron acceptors and they are compensated by a positive space charge formed within the semiconductor. When this happens, the work function of the surface of the semiconductor is roughly independent of the energy of the Fermi level in the interior of the semiconductor. This is represented by the familiar pictorial representation of band bending. When a metal with its own Fermi level contacts the semiconductor surface, equilibrium is reached by electron exchange with the surface states, equilibrating the metal and semiconductor bulk Fermi levels but leaving the space charge unchanged. The development of transistors grew out of this understanding.

Diamond is an insulator with a wide 5.5 eV valence to conduction band gap. Hydrogen-terminated non-oxidized diamond surfaces in vacuum do not conduct. However, when exposed to ambient atmosphere they show conductivity not by electrons but by holes ($h^+$) [4]. It has been proposed that most probably the $h^+$ forms on hydrogen-terminated diamond surfaces from reaction of $O_2$, $H_2O$, and $CO_2$ in the air to form an electron acceptor layer at the interface [5]:

$$4e^-_{\mathrm{dia}} + O_{2,\,\mathrm{air}} + 2H_2O_{\mathrm{air}} + 4CO_{2,\,\mathrm{air}} \rightleftharpoons 4h^+_{\mathrm{dia}} + 4HCO^-_{3,\mathrm{aq}} \quad (7.1)$$

The surface hole concentration at equilibrium is around $10^{13}$ $\mathrm{cm}^{-1}$ [6]. The solvated bicarbonate anions will be electrostatically attracted to the surface and will remain close to it.

The $h^+_{dia} + HCO^-_{3,\,aq}$ form a double layer with no net charge in the interfacial system. The holes are mobile and conducting whereas the bicarbonate anions are held more or less in place by the water network structure. They are analogous to Bardeen's surface states. The $h^+$ are distributed in a sub-surface state space charge region as in Bardeen's description and have sufficient density to allow participation in electrical conduction. The double layer system has a workfunction that is subject to measurements similar to the emersed electrodes with intact double layers discussed in Chap. 5. Might the surface workfunction and hole state occupation be changed by modifying the electrochemical potential of the water solution? Chakrapani et al. addressed this question [7]. As discussed in this paper, from Mott-Schottky measurements the Fermi level for equilibrated diamond surfaces in air is $E_F \approx -5.2$ eV. How can this electrochemical potential be changed?

From the Nernst equation, which is discussed in detail in Chap. 12, for aqueous solutions that are in equilibrium with ambient 0.21 bar atmospheric $O_2$(g), the chemical potential of electrons is dependent of the solution pH according to the following formula for the four-electron reduction to $H_2O$(aq):

$$\mu_e = -4.43\,\text{eV} - 1.229\,\text{eV} + 0.0592\,\text{pH} \tag{7.2}$$

By this equation, the chemical potential of electrons in aqueous solution is $\approx -5.3$ eV when the solution is in equilibrium with ambient atmosphere, for which the pH of approximately 6. Solutions with different pH values will have different potentials and solutions can be prepared with potentials above and below the Fermi level of the hydrogen terminated diamond surfaces exposed to ambient conditions in the lab. For example, the oxygen half-cell potentials under these conditions give electron chemical potentials of $-4.83$ eV and $-5.66$ eV at respective pH values 14 and 0. As shown in Fig. 7.1, these span the diamond Fermi level of $\approx -5.2$ eV for hydrogen terminated diamond exposed to the ambient. When adding high surface area diamond powders to small quantities of solution at pH values corresponding to potentials above and below the $-5.2$ eV level, one expects that there will be electron transfer until a new equilibrium is reached with concomitant change the solution pH values.

Chakrapani et al. prepared H-terminated diamond powders and added 1 g to 20 ml of pH = 5.2 acidic aqueous solution. The electron potential of the solution is $-5.35$ eV, causing electrons to transfer from the diamond to $H^+$(aq), causing the pH to jump up to 8.6 (Fig. 7.2). When the same preparation was added to pH = 8.5 basic solution with electron potential 5.16 eV, electrons transferred from solution to the diamond and consumed $OH^-$(aq), so that the solution became less basic and the pH jumped down to 7.2. The energy diagram in Fig. 7.1 provides prospective connecting the oxygen half-cell reactions, the electronic structures of the diamond and solutions and the pH changes. The pH changes are large enough to be measured when the surface areas of the powder are high and the quantity of the solution to which they are added solution is small.

After the pH equilibrates, following adding diamond powders to the solutions, the new pH values correspond to new electron chemical potentials in the solutions.

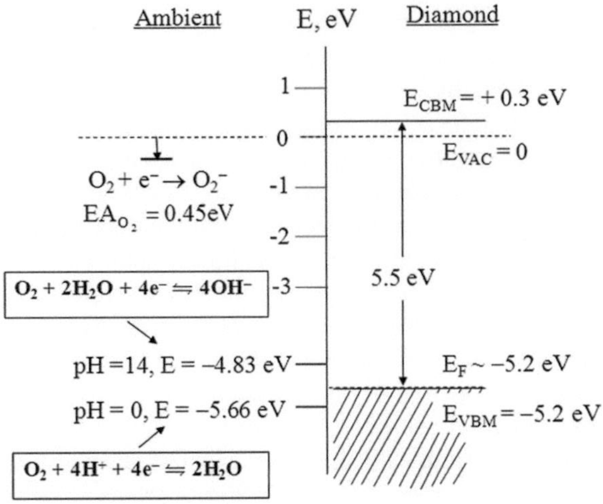


**Fig. 7.1** Schematic electronic structure of diamond in the right and standard oxygen reduction energies in vacuum (−0.45 eV), pH 14 solution (−4.83 eV) and pH 0 solution (−5.66 eV) on the left. The diamond band gap is 5.5 eV and the Fermi energy and valence band maximum are at −5.2 eV. The figure omits the diamond upward valence band bending which accompanied the hole formation at the top of the valence band. The figure shows the Fermi energy determined from Mott-Schottky measurements in the literature rather than the −5.3 eV value from the Nernst equation for pH = 6. From V. Chakrapani, J. C. Angus, A. B. Anderson, S. D. Wolter, B. R. Stoner, and G. U. Sumanasekera, "Charge Transfer Equilibria Between Diamond and an Aqueous Oxygen Electrochemical Redox Couple," Science 318, 1424-1430, 2007. Reprinted with permission from AAAS

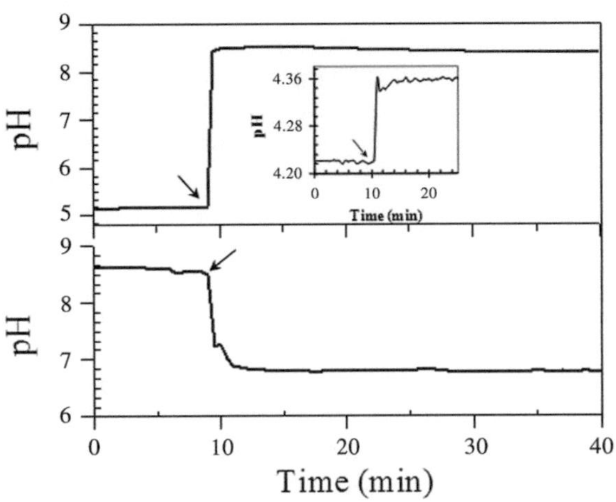


**Fig. 7.2** X from Chakrapani et al. Arrows show times when 1 g hydrogenated diamond powders were added to 20 ml aqueous solutions of initial pH 5.2 and 8.5. The inset shows pH change with deoxygenated solution, which suppresses the oxygen half-cell reaction. From V. Chakrapani, J. C. Angus, A. B. Anderson, S. D. Wolter, B. R. Stoner, and G. U. Sumanasekera, "Charge Transfer Equilibria Between Diamond and as Aqueous Oxygen Electrochemical Redox Couple," Science 318, 1424-1430, 2007. Reprinted with permission from AAAS

Being in equilibrium with the diamond surfaces, it means the chemical potentials of electrons in the surfaces match the solution potentials. The diamond valence band bends up due to $h^+$ formed in the surface region. Compared to the initial maximum of the bent valence band at $\approx -5.2$ eV, the surface band bent up an additional 0.02 eV in the acidic solution due to the additional $h^+$ that are created and it bent down 0.05 eV in the basic one because of $h^+$ depletion.

Hydrogen terminated diamond surfaces exposed to the ambient atmosphere have been the basis of field effect transistors and are discussed in a review by Crawford et al [8]. In this comprehensive review other means of transfer doping and related topics are included. The authors believe that continuing research will "likely deliver a range of diamond-based electronic, quantum and sensing device solutions in the near future."

## 7.3 Solution Electron Potential and Occupation of Mid-Gap States

Returning to the effects of pH on wide band gap insulators, many such oxides have empty mid-gap states. Luminescence can occur when electrons are excited by uv light to the conduction band and then deexcite to the band gap state. If the mid-gap density of states is spanned in whole or in part by the electrochemical potential range of electrons over the pH range, then their occupation can be changed by varying the pH of a solution introduced at the surface and this will affect the color and intensity of luminescence. These effects were observed for gallium nitride and zinc oxide by Chakrapani et al [9]. Figure 7.3 illustrates changing occupation of mid-gap states due to pH and the effect on luminescence near the semiconductor surface.

In summary, the electron chemical potential of a solution interfaced with an insulating solid may be able to induce a double layer with surface conductivity in the insulator by $h^+$ charge carriers at the top of the valence band. If there are no acceptor states near the top of the valence band of the insulator, inducing electron conductivity will not be possible. However, in materials with empty mid band gap states the solution may transfer electrons into the gap states and change energies of luminescence generated by excitation from the valence band to a higher empty band which is followed by deexcitation to the mid band states. In this case if the mid-gap states are partially occupied, electron withdrawal from them may occur and also change the luminescence. Variables controlling these effects include the valence band energies, energies of mid-gap states, solution electron chemical potential parameters, such as pH, and limits imposed on the extent of electron transfer by properties of the solid and solution.

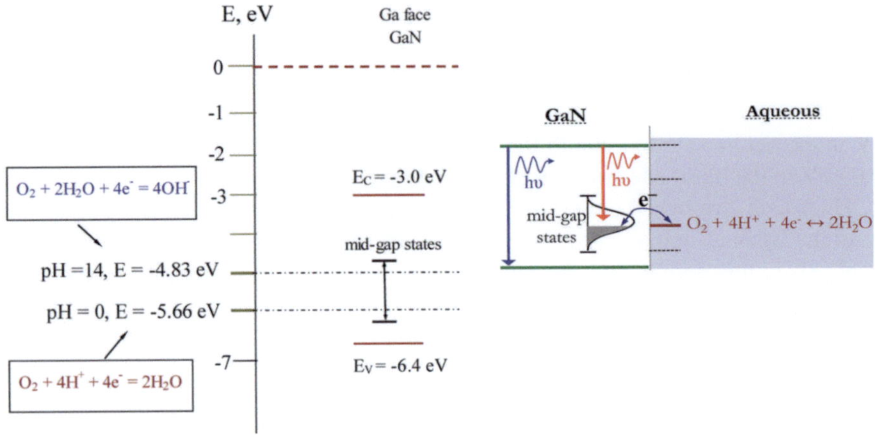


**Fig. 7.3** As in Fig. but for gallium nitride in place of diamond. The conduction band edge is at −3.0 eV and the valence band top is at −6.4 eV. Mid-gap energy states spanning the spanning the 0.83 eV oxygen reduction range are indicated. It is illustrated how the mid-gap states, represented by a Gaussian function, may have variable occupation depending on pH of the electrolyte. Adapted with permission from V. Chakrapani, C. Peddyaia, K. Nash, A. B. Anderson, M. K. Sunkara, and J. C. Angus, "Electrochemical pinning of the Fermi level: mediation of photoluminescence from gallium nitride and zinc oxide," J. Am. Chem. Soc.130, 12944-12952, 2008. Copyright 2008 American Chemical Society

## References

1. Hansen WN, Wang CL, Humphries TW (1978) Electrode emersion and the double layer. J Electroanal Chem 90:137–141
2. Kötz ER, Neff H, Müller K (1986) A UPS, XPS, and workfunction study of emersed silver platinum and gold electrodes. J Electroanal Chem 215:331–344
3. Bardeen J (1947) Forming a double layer surface states and rectification at a metal semiconductor contact. Phys Rev 71:717. https://doi.org/10.1103/PhysRev.71.717
4. Landstrass MI, Ravi KV (1989) Resistivity of chemical vapor deposited diamond films. Appl Phys Lett 55:975–977
5. Chakrapani V, Eaton SC, Anderson AB, Tabib-Azar M, Angus JC (2005) Studies of adsorbate-induced conductance of diamond surfaces. Electrochem Solid State Lett 8:E4–E8
6. Hayashi K, Yanamaka S, Okushi H, Kajumura K (1996) Appl Lett 68:376
7. Chakrapani V, Angus JC, Anderson AB, Wolter SD, Stoner BR, Sumanasekera GU (2007) Charge transfer equilibria between diamond and an aqueous electrochemical redox couple. Science 318:1424–1430
8. Crawford KG, Maini I, Macdonald DA, Moran DAJ Surface transfer doping of diamond: a review. https://doi.org/10.1016/j.progsurf.2021.100613
9. Chakrapani V, Peddyaia C, Nash K, Anderson AB, Sunkara MK, Angus JC (2008) Electrochemical pinning of the Fermi level: mediation of photoluminescence from gallium nitride and zinc oxide. J Am Chem Soc 130:12944–12952. https://doi.org/10.1021/ja710999r

# Chapter 8
# Quantum Theories for Predicting Reversible Potentials and Activation Energies for Electron Transfer Reactions, But Still Omitting Surface Charging and Double Layer Polarization: 1980s to Present Time

**Abstract** This chapter explains the widely employed quantum theoretical models, using predominately density functional theory (DFT), that comprise the advancements in computational technique that followed developments of semiempirical theory and applications presented in Chap. 5, with some additional ones presented here related to passive film formation on iron and CO and $H_2O$ reactions on platinum. The DFT approaches of this chapter were used by many researchers and some continue to be used for calculations made at the potential of zero charge (pzc). Included are the local reaction center (LRC) model for estimating electron transfer activation energies and reaction reversible potentials, the linear Gibbs energy relationship (LGER), and computational hydrogen electrode (CHE) models. The focus is on applications to oxygen reduction reactions relevant to hydrogen fuel cells over platinum and carbon catalysts. Additionally, an LGER analysis of methanol oxidation on platinum is outlined and the important role of OH(ads) as the oxidant of the intermediate CO(ads) as an intermediate during carbon dioxide formation is established. It is shown that in cases where the standard solution reversible potentials for the reaction intermediates are available, desorption energies that reaction intermediates must have in the pzc model to make an electrocatalyst ideal can be predicted.

## 8.1 General Introduction

Some computational approaches to electrocatalysis were extensions of the budding theoretical field of surface chemistry at the vacuum interface. They did not try to attempt to control the potential or change the surface charge. Doing so requires a double layer of ions and solvent molecules to respond to keep net zero charge for the electrochemical interface. And some did not apply the Gurney picture of electron transfer where electrons transfer between acceptor/donor levels and the Fermi level of the electrode surface. The local reaction center (LRC) and computational hydrogen electrode (CHE) models incorporate electrode potential into the calculations by having the electron exchange between the surface model and an external level. In this sense they fit the Gurney picture. The potential is determined by the

 
A. Anderson, *Past and Current Models in Electrochemistry*,
https://doi.org/10.1007/978-3-031-97791-6_8

energy of the level on the SHE scale. In the LRC case the structure of the reaction center is varied until resonance with each selected energy level is attained and energies are graphed as functions of potential. From these graphs, transition state structures and activation energies as well as reversible potentials are determined. In CHE applications reactant and product structures are calculated and the energy of the external electron level, and therefor reversible potential, are the values that made the reaction energy zero for reaction at equilibrium. The linear Gibbs energy relationship (LGER) avoids such direct implementation of the Gurney picture by using perturbation theory. LGER takes the standard reversible potential in solution, either known or calculated, and evaluates the standard reaction Gibbs energy from it. Then reactant and product adsorption energies to the electrode surface are calculated and added to the standard solution reaction energy to give the predicted reaction energy and reversible potential for the reaction on the surface. Calculations by all these methods are at the pzc. This chapter demonstrates some of their applications and limitations.

## 8.2 Early Applications of Semiempirical Theory

While a greater understanding of electrochemical vibrational tuning rates has advanced in recent years, the significance of EVTR measurements to electrochemistry is limited because they give little direct information about chemical reaction mechanisms and electron transfers at electrode surfaces. In aqueous electrolytes the potential window available for Faradaic reactions on platinum electrodes is limited to a few tenths of a volt. The window is caused by upd H and $H_2$ evolution at the low potential end and water adsorption and oxidation at the high potential end, which both block Faradaic reactions. EVTR measurements of adsorbed CO and other adsorbed molecules are constrained to potentials within the window because only then are adsorption sites not already occupied.

The semiempirical MO calculations and perturbation MO interpretations introduced in Chap. 5 provide frameworks for understanding EVTR measurements. First-principle DFT computational approaches directed at calculating the EVTR were also mentioned briefly in Chap. 5. The next two sections of this chapter treat some applications of semiempirical modeling representing early stages in the development of the quantum theory for energies of electrochemical reactions and their mechanisms. These applications may be regarded as experimental as an effort to extend the range of the semiempirical band shifting model, known already to yield qualitative potential-dependent trends for vibrational phenomena of adsorbed molecules on electrode surfaces. Following sections also give examples with *ab initio* and first principle computational applications to electrocatalytic reactions but without controlling the Fermi level and potential of the electrode, that is at the pzc, also in the absence of the electrolyte.

## 8.3 $H_2O$ Reactions and CO Oxidation on Pt and Passive Film Formation on Iron

A study of reactions of the $H_2O$ molecule bonded to five iron atom clusters representing the Fe(100) surface made use of the ASED-MO band shift technique [1]. This exploratory adsorption study employed several VSIP energies for iron to model potential changes with one set was assigned to an electrode potential of zero the SHE scale, the assignment being an estimate. The five atom cluster models for Fe(100) are smaller than the 18- and 19-atom clusters used above to represent Pt and Pd electrodes because those studies were done later and made possible with more available computer power.

The goal was to determine a mechanism for passive film formation on Ferrous metals in the aqueous interface. Passive films are oxide coatings that block deeper oxidation of metals. Its structure on iron it is several tenths of an Å thick and is composed of nanocrystals containing defects [2]. Using ASED-MO, activation energies were calculated for reactions of adsorbed $H_3O^+$, $H_2O$, and OH in the absence of solvation over a 4 V potential range defined within the model. $H_2O$(ads) was predicted to dissociate with a gain of stability and with a small activation energy to OH(ads) + H(ads). OH(ads) dissociation to O(ads) + H(ads) had stability gain only at negative potentials, also with small activation energy. $H_3O^+$(ads) was calculated to be unstable relative to $H_2O$(ads) + H(ads). While these and related studies produced qualitatively sensible conclusions based on knowledge about water reactions on iron, they suffer from the limitations of the lack of charge self-consistency in the ASED-MO theory which may exaggerate bonding and charge transfer metrics as the valence band is shifted. Furthermore, accuracy is reduced by the omission of solvation, explicit charging on the surface, and electrolyte. Because of these deficiencies, conclusions about the reaction mechanisms are far from justified and they can only be viewed as a first stab at the problem. Ultimately, treatments using more accurate methods are providing firmly based conclusions. At the time computers were becoming more and more powerful, opening the door for accurate modeling in the future. However, these semiempirical calculations were starting points and their results connected with chemical reasoning through the well-established perturbation MO theory as discussed earlier. Additional applications of ASED-MO theory produced and supported a plausible overall mechanism for passive film formation [3]. This remains a topic for further advanced treatment using density functional theory. A recent take on the topic based on kinetic analysis is available [4].

The desire to understand reaction mechanisms over the electrocatalysts used in fuel cells was a strong driving force when only semiempirical MO calculation approaches were practicable, leading to such studies despite their many inherent approximations. Hydrogen fuel cells were an important area of application, and already there were complications to deal with before addressing the reaction chemistry of the fuel cells. CO is an impurity that even when present at low levels in $H_2$ gas stream adsorbs strongly enough at low potentials on Pt anodes to poison them by blocking $H_2$ adsorption sites. It is removed at potentials greater than around 0.6 V by oxidizing it to $CO_2$(g) but this means that either the anode potential must be cycled positive from time to time, perhaps with the brief introduction of $O_2$ or it must be run at a potential significantly higher than 0 V, thus lowering the cell voltage. Carbon monoxide is a late stage intermediate during alcohol oxidation when it is used as the fuel. Standard reversible potentials for methanol and ethanol oxidation are close to 0.0 V, and the CO(ads) intermediate quickly poisons Pt anodes in alcohol fuel cells. Understanding how CO(ads) is oxidized to $CO_2$, which desorbs, is a significant goal. An early study modeled the surface Pt(111) with relatively large, for the time, two-layer eighteen atom clusters. CO(ads) oxidation to $CO_2$ was explored with ASED-MO band shift calculations [5]. Transition state energies for the reacting CO and OH placed in adjacent 1-fold sites were found to be too high at the modeled 0 V, ½ V and 1 V potentials for the reaction shown in Eq. (8.1):

$$\mathrm{CO\,(ads) + OH\,(ads) \rightarrow COOH\,(ads)} \tag{8.1}$$

Calculations using high coverage ½ ML modeling of CO(ads) or OH(ads) gave lower activation energies because of weakened adsorption of CO(ads) from through-bond repulsions. The COOH(ads) was assumed to deprotonate to $CO_2$, which desorbed. Activation energies for forming OH(ads) by dissociation of $H_2O$(ads) by the reaction

$$\mathrm{H_2O\,(ads) \rightarrow OH\,(ads) + H\,(ads)} \tag{8.2}$$

were calculated to decrease from 0.44 eV at 0 V to 0.17 eV at 1 V. This led the authors to suggest that at some small positive potential OH(ads) formed and removed the CO poison by Eq. (8.1). The H(ads) formed by Eq. (8.2) is oxidized at potentials of interest, greater than 0.0 V. These theoretical predictions agreed with experimental electrochemical interpretations at the time, though it is argued later that OH(ads) is formed by oxidation of $H_2O$(ads) in acid by the mechanism.

$$\mathrm{H_2O\,(ads) + H_2O \rightarrow OH\,(ads) + H_3O^+ + e^-} \tag{8.3}$$

More recent work used first principle calculations and it was assumed the reaction in Eq. (8.1) removed CO(ads). During methanol oxidation in basic electrolyte during methanol fuel cell operation CO(ads) is an a known intermediate [6] Proposed mechanisms of methanol oxidation will be treated later where Eq. (8.1) will again participate. intermediate [6] Proposed mechanisms of methanol oxidation will be treated later where Eq. (8.1) will again participate.

## 8.4 Influence on CO Oxidation from Alloying Pt with Electropositive Atoms

Alloying platinum with ruthenium, which is relatively electropositive, has long been known to activate it toward methanol oxidation on fuel cell anodes [7]. Using a cluster model with a substitutional Ru atom in the surface layer, ASED-MO calculations predicted an $H_2O$ molecule adsorbed more strongly on the Ru site and dissociated by Eq. (8.2) with about half the activation energy for a Pt site [8]. This finding motivated systematic investigation of 42 alloying atoms for activating OH(ads) formation and releasing to oxidize CO(ads) by Eq. (8.1) [9]. This model study predicted that the relatively electropositive early first and second transition series metal atoms should be active in attracting and dissociating $H_2O$ molecules when substituting for a Pt atom in the Pt surface.

The cathode in fuel cells is the electrode where the oxidant $O_2$ is reduced, the standard reversible potential being $U^0 = 1.229\,\text{V}$ (SHE). Evidence was accumulating that Pt alloys containing electropositive atoms have lower overpotentials than pure platinum for $O_2$ reduction to $H_2O$ [10]. The OH(ads) was thought to be an intermediate during $O_2$ reduction and it blocked sites interfering with $O_2$ adsorption. Alloying the Pt electrode was believed to shift the removal of OH(ads) at to potential, closer to 1.229 V. Increasing the reversible potential for the equilibrium reaction

$$\text{OH (ads)} + \text{H}^+\text{(aq)} + \text{e}^- \rightleftharpoons \text{H}_2\text{O(l)} \tag{8.4}$$

will accomplish this if OH bonds more weakly to the alloyed surface. H(ads) formed by Eq. (8.2) will not play a role in removing OH(ads) by reacting with it because, as shown later, this reaction does not occur at such high potentials and any H present is oxidized from the surface.

Alloy cathodes such as $Pt_3Cr$ were found to undergo atom segregation in the first several surface layers leaving in nearly pure Pt surface skins [11]. First principles band theory calculations supported the observations of skin formation [12]. The adsorption of OH on the skin Pt atoms was apparently weakened relative to pure Pt, shifting the equilibrium in Eq. (8.4) to higher potential, closer to the $O_2$ reduction potential. This reduces the overpotential for the cathode if Eq. (8.4) is the rate limiting step, a conclusion supported in a combined experimental and theoretical analysis [13]. The role of a further reaction where OH(ads) is oxidized to O(ads) is introduced later.

The surface reactions Eq. (8.1) and possibly Eq. (8.2) are without electron and proton transfer. They are important to the kinetics because they set up the reactants that do undergo the redox reactions, in these cases oxidation reactions. Their reaction energies must also be considered and this will be addressed later.

## 8.5 Applications of *ab initio* and First Principles Density Functional Theory (DFT) to Reactions on Electrodes. Transition State Predictions from the Local Reaction Center (LRC) Model at the Potential of Zero Charge (pzc) Using Both *ab initio* and Density Functional Calculations

Advancements over the Butler–Volmer and Marcus theories for electron transfer kinetics began around 1998. Koper and Schmickler applied models to inner-Helmholtz-plane redox reactions [14]. Koper and Voth treated reductive adsorption of $X_2$ diatomic molecules including $Cl_2$ by the Butler–Volmer mechanism. They assumed the $Cl_2$ LUMO mixed with surface orbitals to weaken the Cl–Cl bond and combined Marcus–Hush solvent relaxation energy and Morse potentials to obtain the potential energy surface [15]. A transfer coefficient of 0.28 eV/V was calculated. The model does not seem to have pursued further.

A model for calculating electron transfer activation energies without recourse to parameterizing the Butler–Volmer equation was introduced in 1998 [16]. In this model, now called the local reaction center (LRC) theory, an electron source for reduction, or sink for oxidation is included with the redox center, M. In a reduction reaction the electron affinity, *EA*, of M changes when its structure is distorted. For reduction to take place at a chosen potential an electron affinity must be reached that matches the oxidation potential of the donor electron source, D, which is a molecule placed at distance in the calculation. When this is accomplished an electron can transfer and equilibrium defining the transition state is established:

$$\mathrm{M} + \mathrm{D} \rightleftharpoons \mathrm{M}^- + \mathrm{D}^+ \tag{8.5}$$

The process was demonstrated for $H_2$ evolution from diamond surface models by the reaction.

$$\mathrm{H_2O} : \mathrm{H}^+(\mathrm{aq}) + \mathrm{C} - \mathrm{H} + \mathrm{e}^-(U) \rightleftharpoons \mathrm{H_2O} : -\mathrm{H} - \mathrm{H} - \mathrm{C} \rightarrow \mathrm{H_2O} : +\mathrm{H_2} + \mathrm{C}\cdot \tag{8.6}$$

In this equation an O lone pair that coordinates $H^+$ and C–C and C are part of the cluster model of the diamond surface. The electron comes from a donor Li atom added at a distance beyond orbital overlap from the reaction center. The Li atom ionization potential, *IP*(D), was adjusted by coordinating $H_2O$ molecules to it. The potential was assigned by the equation

$$U = (IP(\mathrm{D})/\mathrm{eV} - 4.6)\mathrm{V} \tag{8.7}$$

Here 4.6 V connects the vacuum scale of the calculations with 0 V on the SHE scale corresponding to the 4.6 eV value assumed at the time for thermodynamic

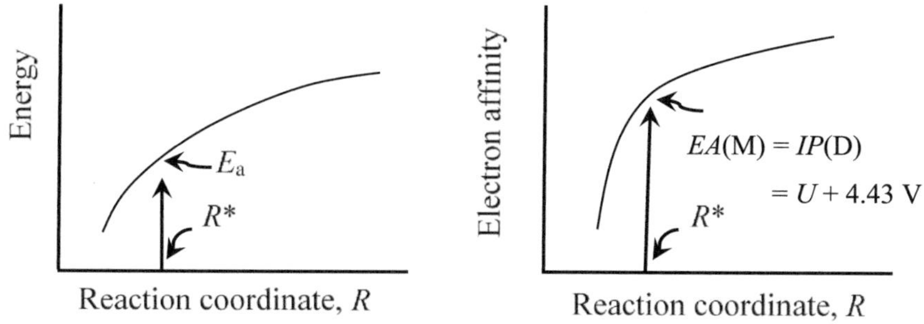


**Fig. 8.1** Relationships between energy, transition state structure, and electron affinity. The left-hand figure is a schematic plot of energy vs reaction coordinate, $R$, with the activation energy marked at the transition state $R^*$. The right-hand figure plots electron affinity vs reaction coordinate and marks the transition state value, which also equals the ionization potential of the donor

workfunction of the standard hydrogen electrode. Nowadays 4.43 eV is known to be a more accurate value. The donor was placed ~10 Å from the reduction center and when placed 20 Å and 50 Å away the results were not significantly altered. Figure 8.1 shows how the energy and electron affinity vary with reaction coordinate and marks the transition state condition. The energy $E$ is the calculated internal energy approximation to the Gibbs energy, $G$. Because the reduction center has more than one structure variable there may be more than one structure with the sought electron affinity, $EA$, and the lowest energy one, $R^*$, gives the activation energy, $E_a$.

*Ab initio* calculations performed with this model were made using the Gaussian 94 code [17]. For some structures the donor-acceptor system went uncontrollably into an excited state following electron transfer. This problem was circumvented by omitting the donor and using the equation, shown here updated to 4.43 in place of 4.6

$$EA(\mathrm{R}) = (U/\mathrm{V} + 4.43)\,\mathrm{eV} \tag{8.8}$$

In the case of oxidation one applies the equation

$$IP(\mathrm{R}) = (U/\mathrm{V} + 4.43)\,\mathrm{eV} \tag{8.9}$$

The LRC theory essentially follows the Gurney model for electron transfer, where transfer occurs when donor and acceptor energy levels are in resonance and is similarly deficient in not including the dependence of bond polarizations on electrode potential. More LRC applications are discussed later.

The ASED-MO band shift method has goals similar to those of the LRC theory but the approach is quite different. As shown above, when CO is adsorbed on a platinum electrode its bond weakens as the electrode potential is dialed in the negative direction because of orbital occupation changes that can be understood using the perturbation theory formulas. With the ASED-MO band shift method the adsorbed reactant orbitals rehybridize and cause polarization changes as the structure changes. Bond forming or breaking reactions may be induced in the calculations, but no electron transfers into or out of the system, which remains at the pzc. The reactant reaches a transition state and the system falls in energy to the product state, the electrode having provided the electron or hole component of the reaction. In terms of band position, the reaction does have potential dependence but there is no current flow. For example, acetylene adsorption and reaction on cluster models of platinum and iron electrodes may have qualitative significance [18]. In the case of platinum, for which experimental electrochemical measurements were available, calculated potential-dependent activation energies for initial steps of acetylene oxidation and reduction were in qualitative agreement. Transition states were found for each band position by pattern searching most relevant structure variables over a coarse grid.

As introduced so far, LRC theory can be viewed as an outer-sphere model for electron transfer activation energies involving an adsorbate-surface cluster model complex. The model for the electrode is a few atoms and adsorption bond polarizations may change during the course of a reaction but the polarizations are determined by structure alone and there is no dependence on the potential for which electron transfer is being modeled. Figure 8.2 illustrates the electron transfer reactions. The donor, D, or acceptor, A, levels are set to a desired potential and structures of the reaction center are searched until the lowest energy one in resonance with the D or A levels are found.

Attempts to represent the surface with larger metal atom clusters showed the potential range for the LRC calculations is limited. A single Pt atom has $IP = 8.99$ eV and $EA = 2.14$ eV. Assuming the workfunction of the SHE is 4.4 eV, this suggests only potentials from around −4.6 V to 2.3 V can be modeled with the reactant bonded to the Pt atom. If the cluster model is increased in size, its $EA$ increases and its $IP$ decreases until the situation is reached where the $EA$ of the cluster by itself controls the electron transfer potential. When M is bonded to the cluster its orbitals

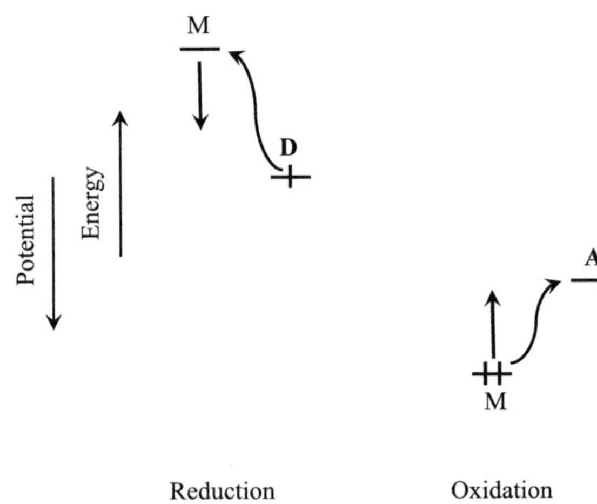


**Fig. 8.2** On the left: schematic showing $EA$(M) increasing during structure strain until it reaches the donor level and transition state for reduction. On the right: $IP$(M) decreases during structure strain until it reaches the acceptor level and transition state for oxidation

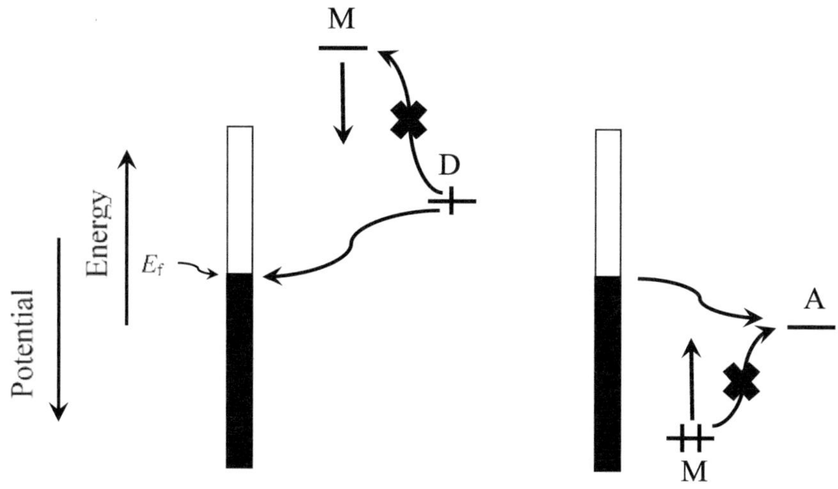


**Fig. 8.3** Schematic figure on the left shows that when, for an electrode, the Fermi level lies below the donor level it is emptied and even if the electron affinity of M increases during structure strain, when it reaches the donor level it is not reduced because the donor level is empty. The figure on the right shows that when the Fermi level lies above an acceptor level the acceptor level becomes occupied and as the ionization potential of M decreases during structure strain, when it reaches the acceptor level it is not oxidized by the acceptor. However, if it rises above the Fermi level it becomes oxidized

are delocalized through the metal atoms, so the variational calculations place the transferred electron or hole in a delocalized orbital. There is still an activation energy to promote the transferred electron up to the orbital associated with bond breaking in M at its transition state. In the bulk limit the cluster *EA* and *IP* converge to the Fermi energy and D will spontaneously give the electron intended for reduction to the Fermi level. Figure 8.3 illustrates the restraints for reduction and oxidation in the bulk surface model limit.

The solution to the problem of the Fermi level stealing the electron or hole from the donor or acceptor at their chosen potentials is to have an adjustable Fermi level for the electrode surface to define the potential rather than an external electron source or sink. How this is done in DFT calculations is described later. First, some examples are given of LRC theory applications to electrocatalysis.

The outer-sphere *ab initio* LRC approach was applied to reduction of $O_2$ to $H_2O$ and oxidation of $H_2O$ to $O_2$ [19]. To get some idea of the catalytic effect of bonding to platinum electrodes, subsequent calculations had the intermediates bonded to a Pt atom [20]. Parameters in the oxidation and reduction precursor structures were optimized by minimizing the energies as functions of structure using the variational principle. For each potential only those structure parameters corresponding to bonds being broken or made and those remaining intact but expected to undergo significant changes in their bond order were included in the optimizations. Three bond lengths were varied when no Pt atom was included and five bond lengths were varied when the Pt atom was included. The remaining bond distances, bond angles, and dihedral angles were kept at the values that were calculated when obtaining structures for the precursors. For reduction reactions the proton donor was modeled by $\left[H - OH_2(OH_2)_2\right]^+$ and for oxidation reactions the proton receiver was the lone pair in $OH_2(OH_2)_2$. The lone pair on the O is a source or sink for protons and the $OH_2(OH_2)_2$ structure was fixed throughout, coming from the optimized structure of $\left[H - OH_2(OH_2)_2\right]^+$. The fixing of structure parameters made the studies practicable but also present another layer of approximation.

The above papers gave explanations for the catalytic effects of forming bonds between intermediates and platinum and connects with the Burton model for stabilizing the reaction intermediates. But as already noted, large cluster models of surfaces are generally not workable for LRC. Besides the absence of potential-dependent polarization changes, the consequences of bond rehybridization that occurs when the potential is changed are absent. Using $OH_2(OH_2)_2$ as a model for the proton accepting water molecule and for the electrolyte is also a simplification. Despite these approximations, the LRC procedure yielded predictions that could be compared with measurements, and provided insights, as discussed below.

A helpful advancement for the computations came with the introduction of constrained variation theory by Kostadinov to automate the calculations of the structures and energies used to predict activation energies [21]. Further improvements were made by others later [22]. Despite all the limitations of the LRC model, it gave helpful insight into redox reaction energetics in those cases where one or two metal atoms were sufficient to describe the bonds between the intermediates and the surface. Examples include H deposition and $H_2$ evolution on one Pt atom, a 1-fold site reaction, and the steps in $O_2$ reduction on two Pt atoms, some steps requiring bridging or 2-fold sites. These are introduced in the next section.

## 8.6 Applications of LRC Model to $O_2$ Reduction and H Deposition on Pt. Potential—Dependent Activation Energies and Reversible Potentials in Acid and Base and Connection to Marcus Normal and Inverted Potential Ranges

The LRC theory can be used to generate $E_a(U)$ for oxidation or reduction steps of molecule R. It can also give reversible potentials, $U_{rev}$, and their activation energies, $E_a(U_{rev})$. The minimum data needed for evaluating a reversible potential are the reduction precursor energy, $E_{rp}$, oxidation precursor energy, $E_{op}$, and the workfunction of the standard hydrogen electrode, taken in early work as 4.6 eV, though 4.43 eV is now known to be more accurate. The formula is

$$U_{rev} = \left(E_{rp} - E_{op}\right)/e - 4.43\,V \tag{8.10}$$

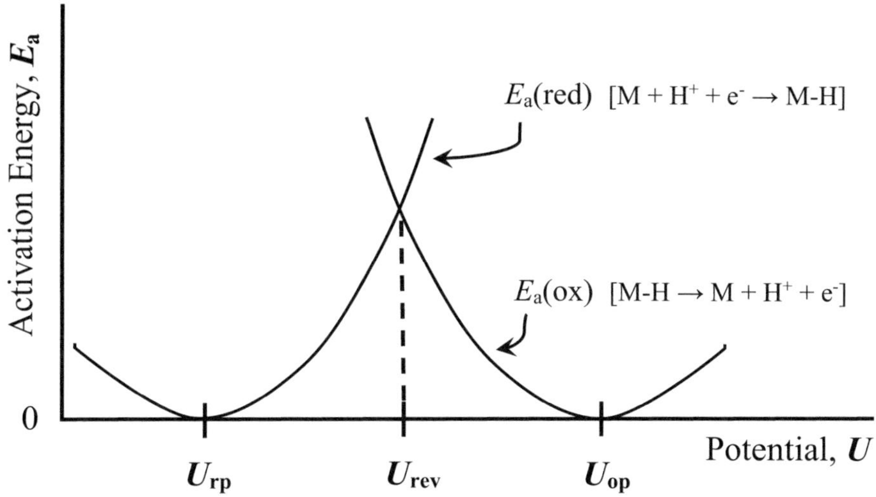


**Fig. 8.4** Relationships between energies and potentials for reduction and oxidation reactions involving proton and electron transfers. M and M–H could be part of a metal surface cluster model, $H^+$ is modeled as $\left[H - OH_2(OH_2)_2\right]^+$, and the electron energy is $[U + 4.43V]eV/V$. The left-hand parabola-like curve has zero reduction activation energy at the reduction precursor potential and similarly for the oxidation activation energy curve on the right which is zero at the oxidation precursor potential. The curves cross at the reversible potential. Reproduced from "How Computational Theorists Deal with Electrode Potential," Alfred B. Anderson 2025 J. Electrochem. Soc. 172 036501 © 2025 Alfred B. Anderson. Published on behalf of The Electrochemical Society by IOP Publishing Limited

Keeping in mind that the calculated internal energy $E$ is an approximation to the Gibbs energy $G$, if the exact Gibbs energies are at hand they could replace $E$ in these equations. The following Fig. 3.4, adopted from Fig. 3.6 of Chap. 3, defines $U_{rp}$, $U_{op}$, $U_{rev}$, $E_a(\text{ox})$, and $E_a(\text{ox})$ for H and $H^+$.

In practice, the curves are drawn through points found as discussed above, each energy obtained from the variational calculations for a series of potential values. In Chap. 3, M referred to a metal atom but from here on we will mostly consider other moieties and replace "M" with "R" as commonly done by organic chemists. The reduction precursor potential $U_{rp}$ is found for the reduction of $RH_{rp}{}^+$ at the condition of equilibrium:

$$RH_{rp}^+ + e^- \rightleftharpoons RH_{rp} \tag{8.11}$$

The energy of the variationally optimized reduction precursor $RH_{rp}^+$ is used in this equation. The energy balance equation at equilibrium is:

$$E\left(RH_{rp}^+\right) = E(RH_{rp}) + 4.43\,\text{eV} + eU_{rp} \tag{8.12}$$

Electron transfer is taken to occur suddenly at all potentials as in the Gurney model so that the structure of RH is unchanged from that of the optimized structure of $RH^+$.

The reduction potential for $RH^+_{rp}$ is $U_{rp}$ and depends on its electron affinity $EA_{rp}$:

$$U_{rp} = \left(EA_{rp} - 4.43\,\text{eV}\right)\text{V/eV} \tag{8.13}$$

Similarly, for oxidation, the energy balance at equilibrium is written:

$$E\left(\text{RH}_{op}\right) = E\left(\text{RH}^+_{op}\right) - 4.43\text{eV} - eU_{op} \tag{8.14}$$

where the oxidation precursor potential, $U_{op}$, depends on its ionization potential, $IP_{op}$:

$$U_{op} = \left(IP_{op} - 4.43\,\text{eV}\right)\text{V/eV} \tag{8.15}$$

Next is shown how activation energies for reduction and oxidation are connected. The activation energy $E_a(\text{red}, U)$ for reduction of $RH^+$ which has optimized structure $RH^{+\prime}$ at potential $U \neq U_{rp}$, is

$$E_a(\text{red}, U) = E\left(\text{RH}^{+\prime}, U\right) - E\left(\text{RH}^+_{rp}\right) \tag{8.16}$$

Similarly, the activation energy for oxidation of site RH at potential $U \neq U_{op}$, is

$$E_a(\text{ox}, U) = E\left(\text{RH}'', U\right) - E\left(\text{RH}_{op}\right) \tag{8.17}$$

At potential $U$ the transition state structure for $RH^{+\prime}$ reduction is identical to the transition state structure for $RH''$ oxidation. Letting RH stand for this structure and writing Eq. (8.14) for potentials $U \neq U_{op}$, the following equation can be written:

$$E(\text{RH}, U) = E\left(\text{RH}^+, U\right) - 4.43\,\text{eV} - eU \tag{8.18}$$

Similarly, from Eq. (8.16),

$$E\left(\text{RH}^+, U\right) = E_a(\text{red}, U) + E\left(\text{RH}^+_{rp}\right) \tag{8.19}$$

Substituting Eq. (8.19) into Eq. (8.18) and putting the resulting equation into Eq. (8.17) without the primes and gives a formula for $E_a(\text{ox}, U)$ in terms of $E_a(\text{red}, U)$, where $RH^+$ and RH are omitted for brevity:

$$E_a(\text{ox}, U) = E_a(\text{red}, U) + E_{rp} - E_{op} - (4.43 + U/\text{V})\text{eV} \tag{8.20}$$

Similar formulas may be written for finding $E_a(\text{red}, U)$ from $E_a(\text{ox}, U)$:

$$E_a(\text{red}, U) = E_a(\text{ox}, U) + E_{\text{op}} - E_{\text{rp}} + (4.43 + U/\text{V})\text{eV} \quad (8.21)$$

The above relationships are depicted in Fig. 8.5. The reversible potential, $U_{\text{rev}}$, corresponds to the potential with $E_a(\text{ox}, U) = E_a(\text{red}, U)$:

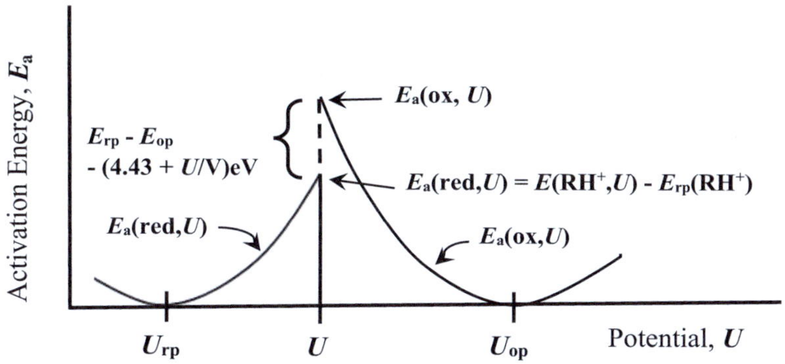


**Fig. 8.5** Two partial non-intersecting parabola-like curves with zeros at the reduction and oxidation precursor potentials showing how, for each potential $U$, when $E_a(\text{red}, U)$ is calculated the value of $E_a(\text{ox}, U)$ may be derived using Eq. (8.20) to generate the oxidation activation energy curve

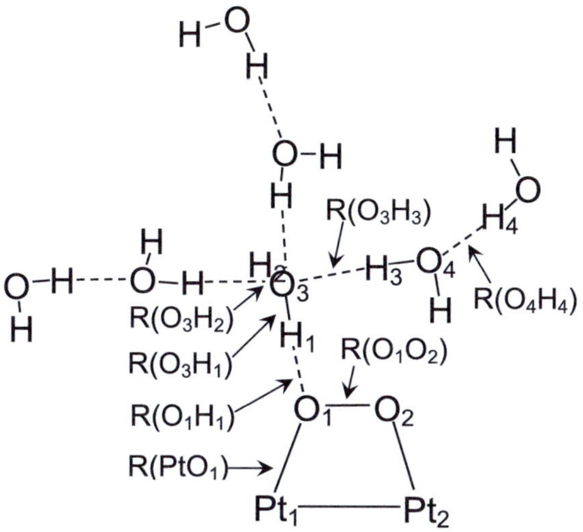


**Fig. 8.6** Showing the seven variationally optimized structure parameters for potential-dependent transition states for forming $OO^-$ from $O_2$ on $Pt_2$. One H of a water molecule is coordinated by hydrogen bonding to $O_2$ shown bridging the Pt atoms in a di-sigma structure. Three water molecules are coordinated through an H atom to the O of the first water and each of these three has a water molecule coordinated through an H to its O atom. Reprinted from Electrochim. Acta 53, T. Zhang and A. B. Anderson, Oxygen reduction on platinum electrodes in base: Theoretical study, 982–989. Copyright 2007, with permission from Elsevier

$$E_{op} - E_{rp} = (4.43 + U_{rev}/V)eV \tag{8.22}$$

As expected, this is the same as Eq. (8.10).

The beauty of equations (8.20)–(8.22) is that only one curve, either $E_a(\text{ox}, U)$ or $E_a(\text{red}, U)$ need be calculated and the other is derivable form it. However, convergence issues when calculating the lowest energy structure for a given potential may complicate applying Eq. (8.21), as discussed in section 8.9. Only small local reaction centers are successful producing energies with these formulas. Reactions on surfaces modeled with band theory will require a different formalism for calculating $E_{rp}$ and $E_{op}$ and activation energies. It will be necessary to change the potential by charging the translational unit cell and including a compensating charge in the electrolyte, thus modeling the potential dependence of the actual electrochemical charge transfer process. Since the Fermi level changes as adsorbate structures change, many calculations and extrapolations will be needed to determine $E_a(\text{red}, U)$ and $E_a(\text{ox}, U)$ and match their structures to find transition state structures and activation energies. Later it is shown how such extrapolations have yielded reversible potentials but for transition states the number of calculations needed may become overwhelming and overcoming this restriction is a research problem.

Reduction activation energies in potential regions to the left of $U_{rp}$ in Figs. 8.4 and 8.5 are in the Marcus "inverted" potential region for the reduction reaction. Oxidation activation to the right of $U_{op}$ are in the "inverted" potential region for oxidation. However, there is no inverted behavior for metal electrode surfaces. This is because when the potential is shifted negative of $U_{rp}$ there is resonance of the acceptor level that was attained for the structure at $U_{rp}$ with occupied band states below $E_f$ and one of them will supply the reducing electron. That is, at lower potentials there is no need to travel up the energy curve negative of $U_{rp}$ in Fig. 8.5 to reach the resonance condition. This means Eq. (8.21) no longer applies for potentials negative of $U_{rp}$ and the curve for reduction activation energy in Fig. 8.5 effectively remains at zero in this range. By the same reasoning, as the potential in increased beyond $U_{op}$ an empty band state will be in resonance to accept the electron from the oxidized reaction center and the activation energy for oxidation effectively remains zero at potentials greater than $U_{op}$. Figure 8.1 in the first chapter may be viewed to qualitatively visualize the densities of states.

Damjanovic, Conway, and Sepa, generated Tafel lines from measurements of $O_2$ reduction in acid at different temperatures. These potential vs log of current density plots at different temperatures intersect at the potential where the activation energy is zero. They called these "critical potentials," $E_{cr}$ [23] See also work on pH effects [24]. The critical potentials are the same as the precursor potentials in the above theory treatment. Damjanovic and Sepa fit Tafel lines defined by measurements of low current density at high overpotentials. Potential vs log of current density plots can show scatter and curvature, making the critical potentials obtained by fitting lines to them and extrapolating to crossing points uncertain. At pH approximately 1.0, $E_{cr}$ was found to be about 0.25 V. The Tafel plots and critical potentials were assumed to be caused by the first electron transfer, forming OOH(ads). Using symmetry factors, the activation enthalpy was about 27 $kJmol^{-1}$ or about 0.26 eV at 0.8 V(RHE), meaning the activation energy remains low in the potential range where current densities for $O_2$ are known to become maximum on platinum electrodes. An explanation of this current limiting behavior is proposed later.

## 8.7 Applying the LRC Model for $O_2$ Reduction to OOH(ads) on Pt in Acid

The LRC model was used to calculate the reduction precursor potentials for $O_2$(ads) reduction to OOH(ads), the adsorption site modeled by $Pt_2$ [25]. This paper contains extensive study of the effects the flexibility given to the reaction center in terms of the choice of structure variables on and the choice of methodology for calculating $E_a(\mathrm{ox}, U)$ and $E_a(\mathrm{red}, U)$ on the predicted precursor potentials and activation energies. Values of about 0.39 V to 0.56 V were obtained, the former for a 7-variable planar reaction center model and the latter for a 10-variable model with a twisted peroxide-like structure. These potentials are higher than the 0.25 V value of $E_{cr}$ from the Tafel plot extrapolations of Damjanovic and Sepa, and the calculated activation energies at 0.55 V higher potentials are, respectively, about 0.3 eV and about 0.2 eV, which are both close to the 0.26 eV value they found. While the approaches could be construed to compare favorably, the uncertainties in the experimental model and extrapolations and particularly the approximations of the theoretical model have yet to be adequately resolved. For example, will hydrogen bonding restraints in the electrolyte limit the twisting? These calculations included an advancement over the original Albu model of LRC which was estimating the field from the electrolyte at the reaction center. The electrolyte was modeled as a rigid array of 0.1 M concentration electrolyte cations and anions using the NaCl structure and the Coulomb contributions at the reaction center were summed to give the Madelung potential[26] and this was added to the Hamiltonian. The Madelung potential put the theoretical potential scale in rough agreement with experiment. The electrolyte charge model is a primitive approximation but it is consistent the absence of surface charging and double layer formation so that it may be viewed as a zeroth order approximation to including the electrolyte.

## 8.8 Introducing pH to the LRC Model Via the Nernst Equation to Calculate $O_2^-$ Formation on Pt in Base

Using Eq. (8.20) to determine $E_a(\mathrm{ox}, U)$ saves the effort of calculating the $E_a(\mathrm{ox}, U)$ curve using Eq. (8.17). In practice not all bond lengths and angles in the LRC model are varied. There are many local minima in the energy that may trap the variational energy calculations in structures with energies above the true minimum energy at each potential. In many cases there two or three variables that change the most in the transition state structures and they correspond to large changes in bond orders during the reaction. A study of oxygen reduction in base is illustrative [27]. Figures 8.6, 8.7 and 8.8 show the transition state structure variables in the LRC calculations, their potential-dependent values, and their activation energies.

Figure 8.8 has the potential scale labeled in SHE and in the RHE scale at pH 13. The relationship between scales uses the Nernst equation: $U_{\mathrm{SHE}} = U_{\mathrm{RHE}} - 0.0592\mathrm{pH}$. The activation energy is only 0.08 eV at the −0.29 V(SHE) reversible potential. Since the model does not take into account effects electrode surface charging and double layer polarization, the predictions are rough. Nevertheless, it turns out that the predicted activation energy for reduction at 0.35 eV overpotential, as shown in Fig. 8.8, is in the range of quoted measured values, 0.38 eV to 0.49 eV, from Arrhenius plots for (100), (110), and (111) [28]

According to Fig. 8.8, the reduction precursor corresponds to the water molecule H2O3H1 hydrogen (Fig. 8.6) bonded to adsorbed $O_2$ which upon electron transfer becomes adsorbed superoxide, $O_2^-$(ads), with the hydrated water molecule coordinated to it. In Fig. 8.7, the OO bond lengthens in the oxidation precursor as the potential increases and the water molecule moves closer. The $O_2^-$(ads) was estimated to dissociate with low activation energy to $O^-$(ads) and O(ads). The $O^-$(ads) abstracted a proton from $H_2O$, forming OH(ads) and $OH^-$(aq). The activation energy for this was about 0.3 eV.

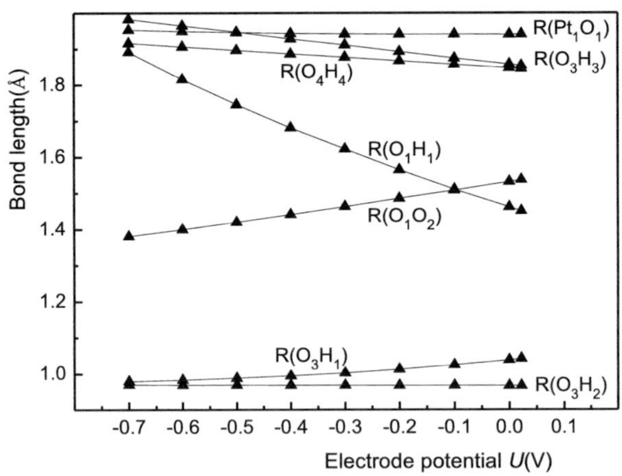


**Fig. 8.7** Optimized structure parameters defined in Fig. 8.6 for the electron transfer transition states for forming and oxidizing $O_2^-$(ads) plotted over 0.1 V steps in the potential range ~ −0.7 V to ~ 0.0 V (SHE) in base. As the H–O bond in the water molecule is stretched, the H–OO hydrogen bond decreases from 1.9 Å hydrogen bond at −0.7 V to about 1.5 Å at 0.0 V. At the same time the O–O bond stretches from about 1.4 Å to more than 1.5 Å. The other internuclear distances change relatively little. Reprinted from Electrochim. Acta 53, T. Zhang and A. B. Anderson, Oxygen reduction on platinum electrodes in base: Theoretical study, 982–989. Copyright 2007, with permission from Elsevier

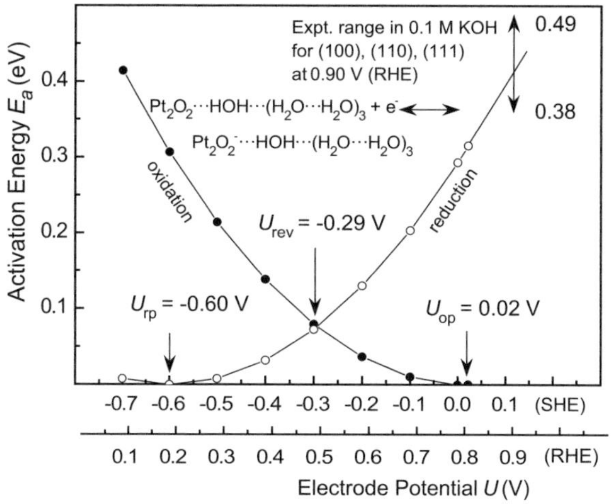


**Fig. 8.8** Potential dependencies of activation energies calculated as functions of electrode potential for forming and oxidizing $O_2^-$(ads) in base. Structure and structure parameters are in Figs. 8.6 and 8.7. The curve rising from the reduction precursor at −0.6 V crosses the curve rising from the oxidation precursor at 0.02 V. They cross at −0.29 V, which is the predicted reversible potential on the SHE scale. The reduction curve passes between experimental results of 0.38 eV and 0.49 eV at 0.1 V in 0.10 M KOH. Solid dots are calculated results and open circles are derived from them. For comparative purposes, shown beneath the SHE scale is the RHE scale for pH 13. Reprinted from Electrochim. Acta 53, T. Zhang and A. B. Anderson, Oxygen reduction on platinum electrodes in base: Theoretical study, 982–989. Copyright 2007, with permission from Elsevier

Comparing the potential-dependent structure parameter curves in Fig. 8.7 with the potential-dependent activation energy curves in Fig. 8.8, the internuclear distances are seen to extrapolate smoothly into the Marcus forbidden potential regions. This is a forbidden region when the reaction complex participates in outer-sphere reduction, but as discussed above, on a metal surface the activation energy is zero when the potential is less than $U_{rp}$. To understand the relationship between structure changes and the Marcus forbidden region in this model, consider a potential point in the inverted region. For it the *EA* of the LRC is fixed to a value that is less than the *EA* of the reduction precursor and, by means of constrained variation theory, the lowest energy structure with this *EA* is calculated. To achieve this value for *EA*, the OO internuclear distance decreases and the H associated with the reduction moves away from the $O_2$. Both changes cost energy which causes the activation energy curve to rise. The shortening of the $O_2$ bond pushes the acceptor $\pi^*$ higher in energy, until the chosen EA is achieved. The other five structure parameters change by small amounts but contribute to minimizing the LRC energy while achieving the chosen EA.

## 8.9 Steps Reducing O to OH and OH to $H_2O$ on Pt in Base

The LRC model for reducing $Pt_2O \cdots HOH$ to $Pt_2OH \cdots OH^-$ had six structure variables, the same as in the previous section without the O–O bond. Activation energy curves for reduction and oxidation may be seen in Fig. 8.9 and they show good agreement between calculated and derived energies, though the crossing point at −0.24 V differs slightly from −0.21 V determined from the precursor potentials.

For the last step, $Pt_2OH \cdots HOH$ reduction to $Pt_2OH_2 \cdots OH^-$ seven structure parameters were again varied. Activation energy calculations showed significant convergence errors which are evident in Fig. 8.10 where the calculated and derived activation energy curves differ by several hundredths of an eV in places. The reversible potentials determined from the crossing points of the activation energy curves lie far from the average of the calculated values using $U_{rp}$ and $U_{op}$, which is about 0.1 V. The asymmetry is the consequence of an inconsistency in the number of structure variables that are variationally optimized when calculating precursor structures and the structures for the $E_a$ curves. For the precursor structures the energies are minimized with respect to all atom positions. However, to make the calculations tractable, the activation energies at potentials less than and greater than the precursor potential were calculated by varying a partial set of structure variables within the constrained variational framework and keeping the rest at the original optimized values. Both methods gave the same reversible potential in Fig. 8.8 and nearly the same value in Fig. 8.9, but Fig. 8.10 shows how there can be large disagreements.

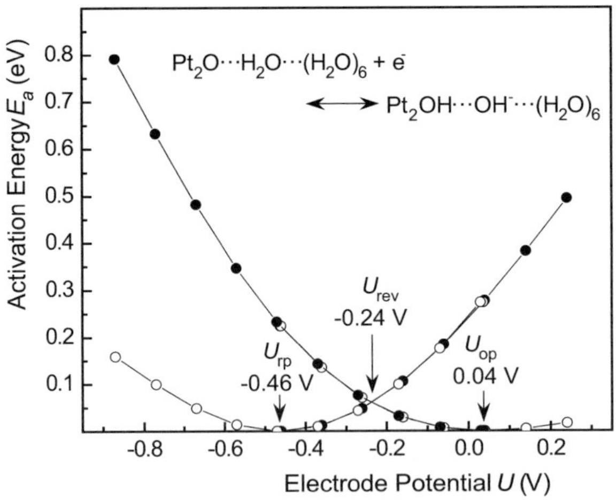


**Fig. 8.9** Potential dependencies of activation energies calculated for forming and oxidizing OH(ads) on a $Pt_2$ bridge site in base. Solid dots are calculated results and open circles are derived results and they lie on the same curve. The reduction and oxidation precursor potentials are −0.46 and 0.04 V, respectively. The reversible potential is −0.24 V with activation energy 0.06 eV. Reprinted from Electrochim. Acta 53, T. Zhang and A. B. Anderson, Oxygen reduction on platinum electrodes in base: Theoretical study, 982–989. Copyright 2007, with permission from Elsevier

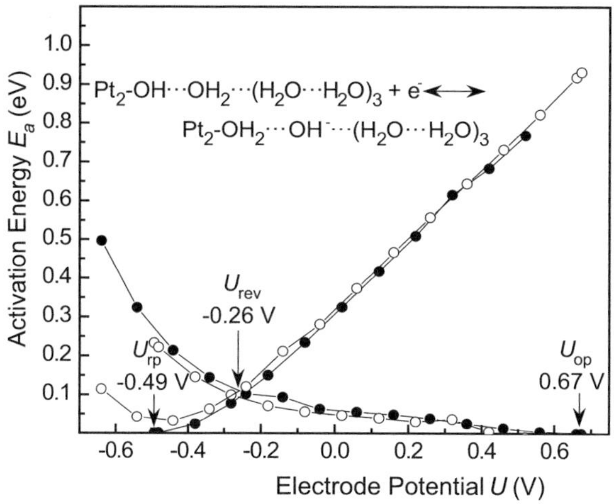


**Fig. 8.10** Potential dependencies of activation energies calculated for forming and oxidizing $H_2O$(ads) on a bridging site in base. Solid dots are calculated results and open circles are derived results. The reduction predursor potential is −0.49 V and for oxidation it is 0.67 V. The curves cross at about −0.26 V with activation energy about 0.1 eV. Reprinted from Electrochim. Acta 53, T. Zhang and A. B. Anderson, Oxygen reduction on platinum electrodes in base: Theoretical study, 982–989. Copyright 2007, with permission from Elsevier

Accuracy of reversible potentials predicted using Eq. (8.10) may be improved by correcting for adsorption bond strength errors. However, the necessary use of small clusters of metal atoms to represent the electrode surface contributed to the abandonment of LRC calculations of activation energies for electron transfers for molecules adsorbed on electrode surfaces. The most significant takeaway from the studies, which spanned the period 1998 to 2007, was the finding that activation energies for proton + electron transfer were no more than a few tenths of an eV. This implied that during the multistep $O_2$ reduction reaction over catalytic Pt and Pt alloy electrodes the overpotential might be determined by the step with the lowest reversible potential. This is sometimes expressed by saying the reaction rate is controlled by thermodynamics more than kinetics. However, another idea, that of effective reversible potential, also a thermodynamic argument presented in Chap. 11 applies and is more general.

Advancements in computer speed since 2007 should make it worthwhile now to pursue LRC calculations with full structure optimization at all electron potentials. Thermal contributions might be calculated and added to the calculated internal energies to give Gibbs energies. However, for reasons given earlier, the method will still not be applicable for large metal atom cluster models or surfaces represented by band theory. In the case of band calculations, advanced DFT simulations are being explored and they will be mentioned later.

## 8.10 Perturbing Standard Values to Predict Reversible Potentials for Catalyzed Electron Transfer Reactions: Linear Gibbs Energy Relationship (LGER) Approach

It would be a useful advancement to find a way to predict reversible potentials for electrocatalytic reactions by means of modifying or extrapolating from experimental tabulations of standard values or perhaps by using theoretically predicted standard values in their place. The standard potentials, measured for standard conditions of activity, pressure, and temperature, already incorporate the consequences of solvation and reaction and product structure relaxations. That is, having them in hand means these difficulties have already been taken into account and "solved" exactly for solution reactions without performing quantum mechanical simulation. The quality of the experimental values is usually high, with several digits past the decimal point reported. When a reactant or product molecule is bonded to the catalyst, part of its solvation shell is replaced by the bond. When this is the case for both reactant and product, solvation changes from the bulk may be thought to not to be small perturbations to theoretical simulations being performed without including solvation. The logic of chemistry makes use of perturbations of known systems as in, for example, substituent inductive effects. What might be an appropriate perturbation here? It could be the bonds to the catalyst replacing part of the solvation shell. Figure 8.11 illustrates the concept [29].

Because reversible potentials for electron transfer reactions are proportional to the reaction Gibbs energy, the reversible potentials when reactants and products are adsorbed on the electrocatalyst will be shifted to new values due to their adsorption energies. The reversible potential for the reaction when reactants are bonded to the

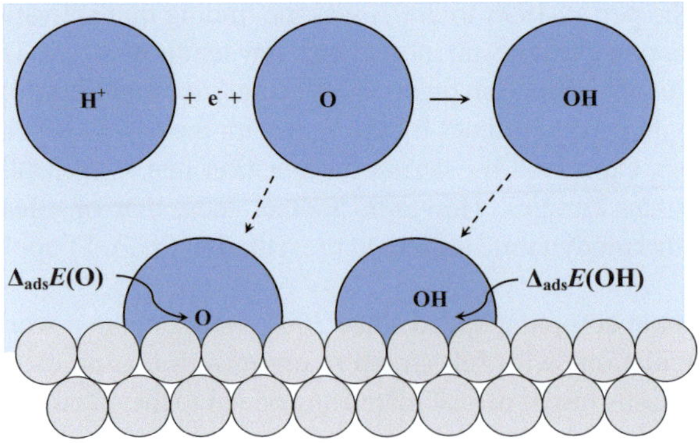


**Fig. 8.11** When the reactants and products solvated in solution adsorb part of the solvation shells are viewed as replaced by bonds to the electrode surface. The figure shows H, O, and OH surrounded by spheres in the bulk solution indicating inclusion of their solvation shells. Then it shows a cross-sectional view of O and OH adsorbed on a Pt(111) slab but now parts of the O and OH solvation shells are replaced by overlap with Pt atoms and the respective adsorption energies are indicated. Reproduced from Ref. [29] with permission from the Royal Society of Chemistry

surface of the electrode, $U_{\text{ads}}$, is

$$U_{\text{ads}} = U^0 + [\Delta_{\text{ads}}G(\text{P}) - \Delta_{\text{ads}}G(\text{R})]/nF \tag{8.23}$$

Here $U^0$ is the standard potential and $\Delta_{\text{ads}}G(\text{P})$ and $\Delta_{\text{ads}}G(\text{R})$ are the respective adsorption energies for the reactant and product. The adsorbed reactant and product are, of course, in their variationally optimized structures. This formula is the basis for an analyses of reversible potential shifts caused by alloying platinum [30]. Figure 8.11 displays internal energies $E$ in place of $G$ because internal energies may be used in Eq. (8.23) while retaining useful predictions of $U_{\text{ads}}$ and they are more easily calculated. Internal energies were substituted for Gibbs energies in the above paper and many subsequent ones.

Mechanistic predictions are possible for non-Pt electrodes too, for example concerning $O_2$ reduction steps on graphite and nitrogen-doped graphite steps [31], cobalt-doped nitride graphite [32], and transition metal chalcogenides [33]. Additional applications may be found in the literature.

## 8.11 Predicting Reversible Potentials by the Computational Hydrogen Electrode (CHE) Approach

Another way to view the problem of evaluating reversible potentials for electrode surface reactions is to leave standard solution values aside and calculate the reaction energy with the electron involved assigned an energy referenced to the SHE workfunction. The energy of the electron that makes the reaction energy zero will then have a value corresponding to the predicted reversible potential for the surface reaction. This is called the computational hydrogen electrode or CHE model.

One can calculate the reaction energy, $\Delta G$, for a reduction reaction modeled simply as discussed so far with no surface charging and double layer polarization, by finding the electron energy that puts the following reaction in equilibrium

$$\text{Ox}_{\text{ads}} + \text{e}^- \rightleftharpoons \text{Red}_{\text{ads}} \tag{8.24}$$

In CHE the electron is in the reaction system and, as in the LRC model, its energy is referenced to the thermodynamic workfunction of the SHE, which corresponds to 0 V, so that at equilibrium the reaction Gibbs reaction energy becomes

$$\Delta G_{\text{ads}} = G(\text{Red}_{\text{ads}}) - G(\text{Ox}_{\text{ads}}) + 4.43\ \text{eV} \tag{8.25}$$

The reversible potential based on Eq. (8.25) is

$$U_{\text{ads}} = -\Delta G_{\text{ads}}/F \tag{8.26}$$

Equations (8.25) and (8.26) comprise the computational hydrogen electrode model introduced by the Nørskov lab [34]. There are many applications of this popular method in the electrochemistry literature.

## 8.12 Comparing the LGER and CHE Models

Taking the equation for the reduction in aqueous solution,

$$\Delta G_{aq} = G(\mathrm{Red}_{aq}) - G(\mathrm{Ox}_{aq}) + 4.43\ \mathrm{eV} \tag{8.27}$$

for which the standard reversible potential is

$$U^0 = -\Delta G_{aq}/F \tag{8.28}$$

and adding the respective adsorption energies for the reactant and product, $\Delta_{ads}G(\mathrm{Ox}_{aq})$ and $\Delta_{ads}G(\mathrm{Red}_{aq})$ to Eq. (8.27) gives the equation

$$\begin{aligned}\Delta G_{ads} &= G(\mathrm{Red}_{aq}) + \Delta_{ads}G(\mathrm{Red}) - G(\mathrm{Ox}_{aq}) - \Delta_{ads}G(\mathrm{Ox}_{aq}) + 4.43\ \mathrm{eV} \\ &= \Delta G_{aq} + \Delta_{ads}G(\mathrm{Red}) - \Delta_{ads}G(\mathrm{Ox})\end{aligned} \tag{8.29}$$

The reversible potential based on Eq. (8.29) is Eq. (8.23) above. In practice the $\Delta G_{aq}$ used in Eq. (8.29) may be experimental but a calculated number may also be used. In LGER as derived, it is an experimental number from Eq. (8.28) but of course a calculated value may be used too. For adsorbed species Eqs. (8.25) and (8.29) are both correct. Equation (8.29) is the LGER model. It was noted that the difference between LGER and CHE predictions of reversible potentials for surface reactions depended on the difference between the experimental $\Delta G_{aq}$ from Eq. (8.28) and the calculated value using Eq. (111) [35]. A later paper explored the relationship in greater detail [36]. The results are expressed in Fig. 8.12.

There has been misunderstanding about the relation of LGER to CHE[37]. In Ref. [37] the authors say the LGER method used OH radical as a reference and is therefore "virtually identical" to CHE but no justification for this statement is provided. In fact, as the above shows, for LGER standard experimental bulk reversible potentials are used as input reference parameters while for CHE the work function of the standard hydrogen electrode is the input reference parameter.

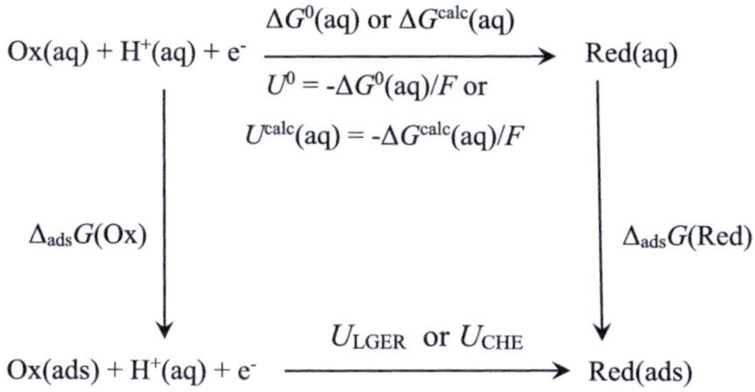


$$U_{LGER} = -[\Delta G^0(aq) + \Delta_{ads}G(Ox) - \Delta_{ads}G(Red)]/F$$

$$U_{CHE} = -[\Delta G^{calc}(aq) + \Delta_{ads}G(Ox) - \Delta_{ads}G(Red)]/F$$

**Fig. 8.12** Starting with the reaction $Ox(aq) + H^+(aq) \rightarrow Red(aq)$, this Haber-like diagram shows that when adsorption energies are added one gets the LGER reversible potential when the experimental $\Delta G^0(aq)$ is used or the CHE result when $\Delta G^{calc}(aq)$ is used, which explains the difference between the two models

## 8.13 Application of CHE to N-Doped Graphene and Energy Staircase Diagram

The energy changes in going through the steps in $O_2$ reduction have been represented graphically in what may be termed "energy staircase" diagrams. These appeared in the 1999 Anderson–Albu paper, in the Nørskov et al. paper and numerous others, including the Anderson–Holby paper from which Fig. 8.13 is taken as an example [38].

The figure is explained as follows:

Along the reaction coordinate, energies at 1 are for $O_2(g)$ + the surface;

The difference between 1 and 2 is the energy released by forming $O_2(ads)$;

the difference between 2 and 3 is the reaction energy released by reducing $O_2(ads)$ to OOH(ads);

the difference between 3 and 4 is the reaction energy released by reducing OOH(ads) to O(ads) + $H_2O$;

the difference between 4 and 5 is the reaction energy released by reducing O(ads) to OH(ads);

the difference between 5 and 6 is the reaction energy released by reducing OH(ads) to $H_2O$.

As seen for the substitutional $FeN_4$ site in doped graphene in Fig. 8.13, it is possible for an intermediate to be energetically trapped as in step 5, meaning energy must be supplied by the surroundings to reach final products. Generally, the figures in the literature show decreasing energies from one reduction step to the next. A monotonic decrease is generally taken as evidence of catalytic behavior, but the numbers themselves quickly reveal whether the catalyst is projected to have useful activity. The Gibbs energy available for $O_2$ reduction to water is 4.92 eV and, for

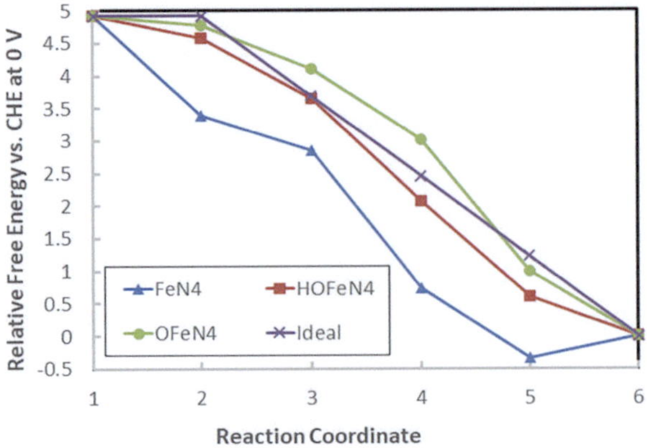


**Fig. 8.13** Energy staircase diagram calculated for $O_2$ reduction steps over iron atom-containing sites in a graphene monolayer. OH and O are bonded to the Fe atom underneath, leaving the top site free for adsorbing reaction intermediates. Reprinted with permission from A. B. Anderson and E. F. Holby, "Pathways for $O_2$ Electroreduction over Substitutional $FeN_4$, $HOFeN_4$, and $OFeN_4$ in Graphene Bulk Sites. Critical Evaluation of Overpotential Predictions Using LGER and CHE Models," J. Phys Chem. 123, 18398–18409, 2019. Copyright 2019 American Chemical Society

ideal behavior each of the four-electron transfer steps must be $4.92/4$ eV $= 1.23$ eV exergonic. Focusing $Fe^{II}$ in the $FeN_4$ site, we see that 1.6 eV energy is lost when $O_2$ adsorbs, and this energy will not be available for electrical work, which this guarantees that the electrical work available is at most 4.9 eV − 1.6 eV = 3.3 eV. However, since energy must be input during the last step to produce $H_2O$ from OH(ads), even less than 3.3 eV will be available. Of course if it were to couple part of the 1.6 eV released during $O_2$ adsorption directly with the last step then the electrical work produced might be greater than 3.3 eV. This idea will not be pursued here.

In the Anderson–Holby paper the O–O bond strengths are calculated to be about 0.8 eV too high. The calculated strengths of O–H bonds when the zero-point bond vibrational energies are omitted are about 0.2 eV greater than when they are included. When calculating the four-electron reduction energy, these overestimates balance 0.8 eV overestimate the OO bond strengths so that the result is close to −4.92 eV. Calculated reversible potentials for the four-electron transfers omitting the zero-point energies are in Table 8.1.

Were the O–O bond strengths 0.8 eV less, correcting for its being overestimated consists in decreasing the energy of the electron participating in the reduction by 0.8 eV so that the reaction is at equilibrium with the products. Were the O–H bond strengths 0.2 eV less, correcting for the absence of zero-point energies the reversible potentials for OOH(ads) formation given in Table 8.2 would each be 0.2 eV less. These corrections add so that the potentials for reduction of OOH(ads) to O(ads) + $H_2O$ would be reduced by 1.0 V. In the subsequent steps the potentials for O(ads) reduction to OH(ads) would be 0.2 V less and the potentials for OH(ads) reduction

**Table 8.1** Reversible potentials, $U_{rev}$ (V), for reactions shown over $Fe^{II}$ , $Fe^{III}$, and $Fe^{IV}$ bulk sites in using the CHE model. $O_2$ adsorption energies are increased 0.13 eV from the calculated values in keeping with the CHE method. The sum of the $O_2$ adsorption energy with $\Delta_{react}\, G^0$ for the four steps is −4.92 eV within round off error at each site. Results in parentheses are outside the $Fe^{IV}$ range of stability. The $Fe^{II}$ site was $FeN_2$ present in the graphene layer and $Fe^{III}$ and $Fe^{IV}$ were created by respectively bonding OH and O to the underside of the $Fe^{II}$ site. Adapted with permission from A. B. Anderson and E. F. Holby, "Pathways for O2 Electroreduction over Substitutional FeN4, HOFeN4, and OFeN4 in Graphene Bulk Sites. Critical Evaluation of Overpotential Predictions Using LGER and CHE Models," J. Phys Chem. 123, 18398–18409, 2019. Copyright 2019 American Chemical Society

| Reaction | Site | | |
|---|---|---|---|
| | $Fe^{II}$ | $Fe^{III}$ | $Fe^{IV}$ |
| | Reversible potential (V) | | |
| $O_2(ads) + H^+(aq) + e^- \rightarrow OOH(ads)$ | 0.53 | 0.94 | (0.68) |
| $OOH(ads) + H^+(aq) + e^- \rightarrow O(ads) + H_2O(l)$ | 2.12 | 1.57 | 1.09 |
| $O(ads) + H^+(aq) + e^- \rightarrow OH(ads)$ | 1.08<br>**potential range of HO – $Fe^{III}$stability** | 1.47 | 2.02 |
| $OH(ads) + H^+(aq) + e^- \rightarrow H_2O(l)$ | −0.34 | 0.60 | (1.00) |

to $H_2O$ would be 0.2 V less. In what follows such corrections do not change the decision about the favored site.

**Table 8.2** First column of numbers is the error in calculated standard reversible potentials; the second contains the CHE prediction of reversible potentials over the $Fe^{III}$ site; and the third has the LGER values. Zero-point energies are not used, requiring a 0.13 eV decrease in the total energy of $O_2$ to make $U^0_{exp} - U^0_{calc}$ zero for the overall four-electron reduction. Adapted with permission from A. B. Anderson and E. F. Holby, "Pathways for $O_2$ Electroreduction over Substitutional $FeN_4$, $HOFeN_4$, and $OFeN_4$ in Graphene Bulk Sites. Critical Evaluation of Overpotential Predictions Using LGER and CHE Models," J. Phys Chem. 123, 18398–18409, 2019. Copyright 2019 American Chemical Society

| Reaction | In bulk solution | On catalyst | |
|---|---|---|---|
| | $U^0_{exp} - U^0_{calc}$ | $U$(CHE) | $U$(LGER) |
| $O_2(g) + 4H^+(aq) + e^- \rightarrow 2H_2O(l)$ | 0.00 | 0.00 | 0.00 |
| $O_2/Fe^{III} + H^+(aq) + e^- \rightarrow OOH/Fe^{III}$ | −0.15 | 0.94 | 0.79 |
| $OOH/Fe^{III} + H^+(aq) + e^- \rightarrow O/Fe^{III} + H_2O(l)$ | 0.75 | 1.57 | 2.32 |
| $O(/Fe^{III} + H^+(aq) + e^- \rightarrow OH/Fe^{III}$ | −0.33 | 1.47 | 1.14 |
| $OH/Fe^{III} + H^+(aq) + e^- \rightarrow H_2O(l)$ | −0.39 | 0.60 | 0.21 |

Here is the interpretation of the table. Beginning with $Fe^{II}$ and $O_2$(ads), the first column of potentials shows $O_2$(ads) will be reduced to OOH(ads) at electrode potentials 0.53 V and below. Might this be the limiting potential for the subsequent reduction steps and therefor the potential at which the electrode operates? No because once OOH(ads) forms, because its reversible potential for reduction is much higher at 2.21 V, it is should be immediately reduced to O(ads) + $H_2O$(l). The adsorbed O atom puts the Fe atom in the IV oxidation state. The O(ads) are reduced at potentials less than 1.08 V to OH(ads), forming $Fe^{III}$, which will be spontaneous when the electrode potential is set to 0.53 V of the first reduction step. The $HO-Fe^{III}$ remains stable to reduction at potentials above −0.34 V but at lower potentials $H_2O$(ads) can form and the site reverts to $Fe^{II}$ again. The CHE prediction is therefor that because of the low potential for the last step the $Fe^{II}$ site will not perform useful catalysis for the four-electron reduction of $O_2$.

Applying this logic to the second column of potentials in Table 8.1, the $Fe^{III}$ site reduces $O_2$ to $H_2O$(l) at 0.60 V and below, corresponding to the OH(ads) reduction, making $Fe^{III}$ a possibly effective catalytic sites for the reaction. The first three reductions have higher reversible potentials, making them allowed at 0.60 V. Steps leading up to OH(ads) and its reduction on $Fe^{III}$ are forming OOH(ads) and then O(ads) which put the Fe in IV and V oxidation states, respectively. From the first column, the range of $Fe^{IV}$ stability is 2.12 V to 1.08 V: Above 2.12 V $Fe^{IV}$ is oxidized and below 1.08 V it is reduced.

Looking at the third column of potentials it is seen that although reversible potentials can be calculated for all the reduction steps over $Fe^{IV}$, the sites are unstable to reduction for the first and fourth reduction steps. This leaves $Fe^{III}$ as the predicted active site with the last reduction step, OH(ads) reduction to $H_2O$(l) having the lowest reversible potential on this site and therefore representing the onset potential for the four-electron reduction. When the $O_2$ and $H_2O$ adsorption energies assumed to be 0.0 eV, the ideal adsorption energies for OOH, O, and OH are 1.35 eV, 2.99 eV, and 1.38 eV, but the calculated values on $Fe^{III}$ were too high by 0.02 eV, 1.00 eV, 1.12 eV. Adding these gives a total energy loss of 2.14 eV and an effective reversible potential 0.70 V for the four-electron reduction, and, in support of the theory, this is close to the 0.60 V value for the OH(ads) reduction step. The effective reversible potential concept will be explained in greater detail in Chap. 11.

If we were to apply the corrections to O–O and O–H bond strengths discussed above by decreasing bond strengths by the amounts of their zero-point energies the reversible potential for OH(ads) reduction over $Fe^{III}$ becomes about 0.4 V. The reversible potentials for the three preceding reductions would remain higher than 0.4 V. One must be wary of taking agreement between a calculated value and a piece of measured data too seriously since calculated numbers may shift around as "corrections" to the model are implemented. It is better to identify a trend in calculated results that can account for what is observed. We know the difference between CHE and LGER, the difference in calculated potentials by the two methods being equal to the errors in CHE calculations of the standard solution phase reversible potentials. Since there are errors in the latter, should we expect LGER predictions to be more accurate?

## 8.14 LGER Results for N-Doped Graphene

As shown in 8.12, $U_{\mathrm{LGER}}$ equals $U_{\mathrm{CHE}}$ + the calculated reversible potential in bulk solution minus the experimentally determined standard value, $U^0$. These differences are given in Table 8.2 alongside the values predicted by CHE and LGER for the $Fe^{III}$ site.

Examining Table 8.2 the expected conclusion that the OH(ads) reduction step sets the limiting potential for LGER as was deduced using CHE. However, the 0.21 V potential for this step, which, applying the zero-point energy "correction," could be reduced to 0.0 V, is far from the 0.7 V onset potential observed over platinum electrodes for $O_2$ reduction. The similarly corrected value for the CHE result is 0.4 V, also far from 0.7 V. Let us again underscore the risk in cherry picking near agreements of predictions of numerical predictions with numerical measurements is assessing the numerical validity of predictions from rough theoretical models. Because of all the approximations made in the calculations one cannot conclude whether CHE or LGER is a better model but we can agree that both predict the last reduction step, the reduction of OH(ads) to $H_2O$, is most likely to be the one determining the overpotential for the four-electron process. Another way to view this is to appeal to the dominance of chemical bond strengths in determining reaction energies and note that changes in the electrode potential, interactions with the electrolyte, variations in hydrogen bonding, and interactions with co-adsorbates are all perturbations to the bond strengths that ultimately must be taken into account for predicting more accurate reaction energies. Until these contributions are taken into account, CHE and LGER predictions are best viewed as suggestions that will motivate further theoretical research. As discussed later, full theories that implicitly include all of these perturbations are a current focus of theoretical development and are improving rapidly. Until the advanced theories are applied to this system it cannot be certain that OH(ads) reduction from $Fe^{III}$ is responsible for the overpotential. Finally, in their review of related experimental literature, Anderson and Holby were not able to identify definitive experimental conclusions about site structure or oxygen reduction mechanisms on $FeN_4$ doped graphite.

## 8.15 LGER Exploration of Six-Electron Methanol Oxidation Based on Calculated $U^0$

The net result of the studies of four-electron reduction of $O_2$ to $H_2O$ using the approximate CHE and LGER models in the previous section was that reducing the OH(ads) intermediate was the lowest potential step and therefor predicted to be the step responsible for an overpotential on substitutional $HOFeN_4$ in graphene. Now an oxidation reaction on platinum is considered. The reactant, methanol, $HOCH_3$ undergoes a six-electron oxidation to $CO_2$ and $2H_2O$. A strongly adsorbed intermediate, carbon monoxide, will form. Recall that CO(ads) in known to poison hydrogen reduction on platinum and in fuel cell applications high purity $H_2$ must be used. The standard reversible potential for methanol oxidation is 0.032 V, very close to the potential of the hydrogen electrode. This means CO(ads) should remain on the surface as it forms from methanol oxidation on an electrode if one attempts to run it at the optimal potential for the highest potential methanol fuel cell. It will be removed by oxidation to $CO_2$ at a potential greater than 0.032 V. Next an analysis is presented of $HOCH_3$ oxidation on Pt(111) based on the LGER model [39].

Experimental values of standard reversible potentials for oxidizing the intermediates were unavailable for this study and so they were calculated using the advanced code called Interface written by Jinnouchi introduced in the next chapter. The adsorption energies were calculated with the same code and the LGER model provided predictions of the reversible potentials for the electron transfer steps on the surface. Two-dimensional slab-band calculations were performed with an 18 Pt atom translation unit cell with three Pt atom layers. Atom spacing initially was from the calculated bulk Pt structure and the top two layers were allowed to relax while the bottom one remained rigid. Adsorbate coverage was 1/6 monolayer (ML), as in Fig. 8.14. This coverage allows only small lateral interactions.

Intermediates in solution and adsorbed are solvated in this application of the Jinnouchi code and energies are calculated at the potential of zero charge. Adsorption energies are in Table 8.3, and they correlate with the degree of unsaturation of the adsorbate and therefor with the number of bonds formed with the surface. A range of shifts to the standard reversible potentials is expected when the intermediates become adsorbed. Structures of the adsorbed intermediates appear in Fig. 8.15.

Table 8.4 has reaction Gibbs energies and reversible potentials calculated for the bulk solution along with reaction energies and predicted reversible potentials for the adsorbed intermediates. The calculated reversible potential for the oxidation is −0.095 V, which is less than the 0.032 V experimental value because of the approximate nature of the theoretical model. The surface reaction energies and reversible potentials have the same sign because these are oxidation reactions. Many of the reversible

**Fig. 8.14** Showing a normal view 1/6 ML methanol coverage on Pt(111). Pt, blue; O, red; H white; C gray. Each O is bonded to a Pt top site and O–H and O–C bonds are an angle to the surface plane. The adsorbed molecules form rows separated by two Pt atoms laterally and by one $Pt_2$ bridge site vertically. Reproduced from Ref. [39] with permission from the Royal Society of Chemistry

**Table 8.3** Most stable adsorption sites and adsorption energies, $\Delta_{ads}E$(eV), for reaction intermediates of methanol oxidation over the Pt (111) surface. The * precedes the atom which forms a bond to the surface. $\Delta_{ads}G$ for formate is for the pzc. Reproduced from Ref. [39] with permission from the Royal Society of Chemistry

| Species | $\Delta_{ads}E$ |
|---|---|
| Methanol (H*$OCH_3$) | −0.277 |
| Hydroxylmethyl (HO*$CH_2$) | −2.131 |
| Hydroxymethylene (HO*CH) | −3.105 |
| Hydroxymethylidyne (HO*C) | −4.442 |
| Carbon monoxide (*CO) | −1.640 |
| Carbon dioxide ($CO_2$) | −0.048 |
| Methoxy (*$OCH_3$) | −1.439 |
| Formaldehyde (*O*$CH_2$) | −0.546 |
| Formyl (O*CH) | −2.392 |
| Hydroxyformyl radical (HOO*C) | −2.320 |
| Formyloxy radical (*O*OCH) | −2.112 |
| Formate anion ($OOCH^-$) | −0.345[a] |
| Formic acid (HOOCH) | −0.111 |
| *$OH_2$ | −0.229 |
| *OH | −1.951 |

potentials are negative and, keeping within the model, since the calculated standard reversible potential for the six-electron oxidation is −0.095 V, these steps might be expected to be spontaneous at this potential but it must be remembered that a saturation coverage of H(ads) is expected at low positive potentials. However, this potential is the goal for a perfect catalyst. The data in Table 8.4 provide several routes for methanol decomposition that lead to CO(ads). Two paths going through OCH(ads) as a precursor to CO(ads) have earlier $H - OCH_n$ bond oxidation steps with high reversible potentials, 0.423V and 0.624 V and the highest potential along a third path is 0.083 V, making it favored of the three. OCH(ads) is oxidized to CO(ads) at a favorable −0.697 V. An HOC(ads) precursor to CO(ads) can also form and it is oxidized to CO(ads) at −0.153 V. Once CO(ads) forms it must be oxidized by reacting with OH(ads) to give the hydroxyformyl radical, HOOC(ads), which is oxidized to $CO_2 + H_2O$ at 0.179 V.

Formaldehyde may form from $OCH_3$(ads), which itself has 0.423 V potential, which makes its formation less likely compared to the above mentioned routes forming CO(ads). When it does form it is bonded strongly enough to the surface with −0.546 eV energy of adsorption that some molecules may be oxidized to OCH(ads) which can be oxidized to CO(ads) as discussed. However, there is a second reaction path where OCH(ads) combines with OH(ads) to give formic acid, HOOCH(ads). The formic acid molecule is weakly adsorbed but it has three viable pathways for oxidation to $CO_2$, through HOOC(ads), $OOCH^-$(ads) and OOCH(ads).

The reversible potential for OH(ads) formation from $H_2O$(ads) is 0.494 V which makes this the probable overpotential producing step for all of the above methanol oxidation mechanisms on platinum. The various pathways are in Fig. 8.16. When

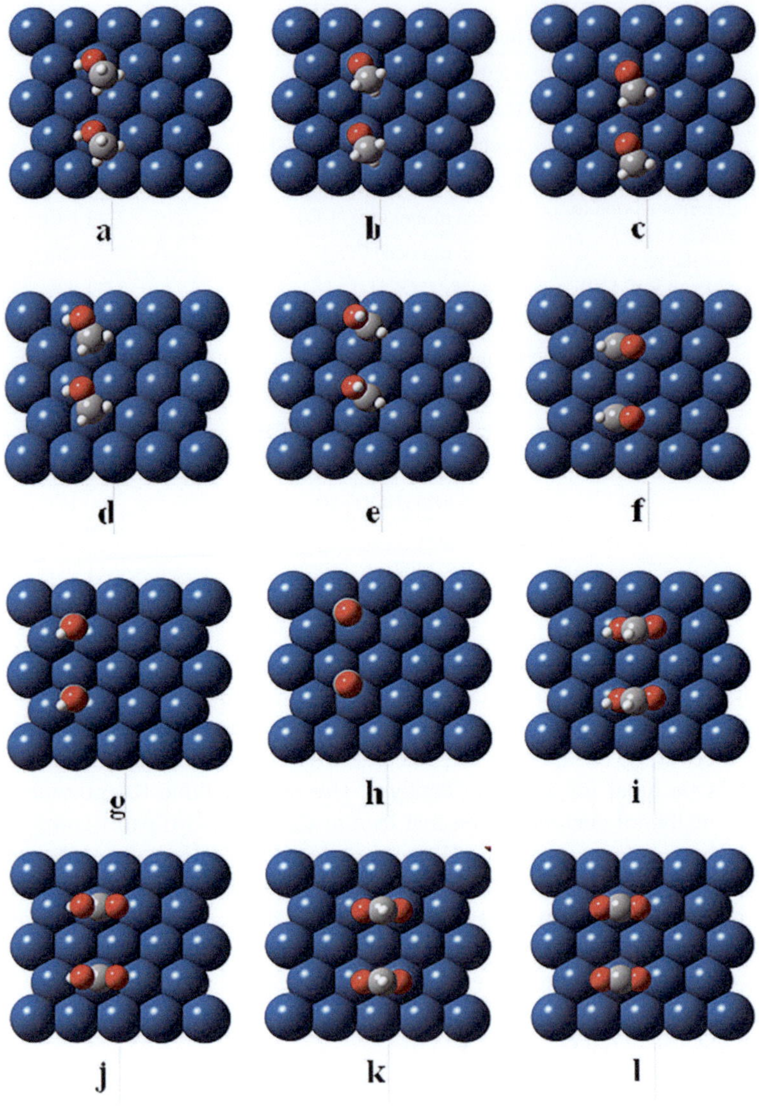

**Fig. 8.15.** Pairs of adsorbed molecules on Pt(111) at 1/6 ML coverage as in Fig. 8.14. **a** methanol, $HOCH_3$; **b** methoxy, $OCH_3$; **c** formaldehyde, $OCH_2$; **d** hydroxylmethyl, $HOCH_2$ ; **e** hydroxylmethylidene, HOCH; **f** formyl, OCH; **g** hydroxymethylidyne, HOC; **h** carbon monoxide, CO; **i** formic acid, HOOCH; **j** hydroxyformyl radical, HOOC; **k** formyloxyl radical or formate anion, $OOCH^-$, and **l** carbon dioxide, $CO_2$. Reproduced from Ref. [39] with permission from the Royal Society of Chemistry

**Table 8.4** $\Delta G^0_{sol}$ is the calculated reaction Gibbs energy for the reactions in solution and $U^0$ is the predicted standard reversible potential. $\Delta E_{surf}$ (eV) is the LGER reaction internal energy on the Pt(111) surface based on the calculated $U^o$(V) and $\Delta_{ads}E$ in Table 8.3. $U^{rev}_{surf}$ is the predicted reversible potential for the surface reaction using $\Delta E_{surf}$ (eV). Also shown are $\Delta G^0_{sol}$ (eV) and $\Delta E_{surf}$ (eV) for reactions for which there is no electron transfer. Reproduced from Ref. [39] with permission from the Royal Society of Chemistry

| Reaction | $\Delta G^o_{sol}$ (eV) | $U^o$(V) | $\Delta E_{surf}$ (eV) | $U^{rev}_{surf}$(V) |
|---|---|---|---|---|
| | −4.43 eV | | −4.43 eV | |
| $HOCH_3 \rightarrow HOCH_2 + H^+(aq) + e^-$ | 1.388 | 1.388 | −0.743 | −0.743 |
| $HOCH_2 \rightarrow HOCH + H^+(aq) + e^-$ | 1.057 | 1.057 | 0.083 | 0.083 |
| $HOCH \rightarrow HOC + H^+(aq) + e^-$ | 0.662 | 0.662 | −0.565 | −0.565 |
| $HOC \rightarrow CO + H^+(aq) + e^-$ | −2.955 | −2.955 | −0.153 | −0.153 |
| $H_2O \rightarrow OH + H^+(aq) + e^-$ | 2.445 | 2.445 | 0.494 | 0.494 |
| $HO + CO \rightarrow HOOC$ | −1.076 | | 0.195 | |
| $HOOC \rightarrow CO_2 + H^+(aq) + e^-$ | −2.093 | −2.093 | 0.178 | 0.178 |
| $HOCH_3 \rightarrow OCH_3 + H^+(aq) + e^-$ | 1.862 | 1.862 | 0.423 | 0.423 |
| $OCH_3 \rightarrow OCH_2 + H^+(aq) + e^-$ | −1.416 | −1.416 | −0.523 | −0.523 |
| $OCH_2 \rightarrow OCH + H^+(aq) + e^-$ | 1.157 | 1.157 | −0.689 | −0.689 |
| $OCH \rightarrow CO + H^+(aq) + e^-$ | −1.449 | −1.449 | −0.697 | −0.697 |
| $HO + OCH \rightarrow HOOCH$ | −4.012 | | 0.190 | |
| $HOOCH \rightarrow HOOC + H^+(aq) + e^-$ | 1.517 | 1.517 | −0.692 | −0.692 |
| $HOCH_2 \rightarrow OCH_2 + H^+(aq) + e^-$ | −0.943 | −0.943 | 0.642 | 0.642 |
| $HOCH \rightarrow OCH + H^+(aq) + e^-$ | −0.843 | −0.843 | −0.130 | −0.130 |
| $HOOCH \rightarrow OOCH + H^+(aq) + e^-$ | 2.035 | 2.035 | 0.034 | 0.034 |
| $OOCH \rightarrow CO_2 + H^+(aq) + e^-$ | −2.612 | −2.612 | −0.548 | −0.548 |
| $HOOCH \rightarrow OOCH^- + H^+(aq)$ | 0.231 | | −0.005 | |
| $OCOH^- \rightarrow OOCH + e^-$ | 1.805 | 1.805 | 0.038 | 0.038 |

following paths around some of the loops in the figure the energies do not always add up to exactly zero. This can be caused by rounding off or the presence of transcription errors in some tabulated numbers.

Experimentally, the onset potential for 1M methanol oxidation on Pt(111) electrodes is about 0.4 V [40]. Interestingly, methanol oxidation voltammograms for the (100) and (110) have different onset potentials and different shapes, indicating structure effects. The measurements show that current densities become high, ~0.4 ma/cm$^2$ around 0.6 V to 0.7 V and continues to increase on these surfaces as the potential is further increased. The problem for fuel cell application is that these potentials have run into the working potential of the fuel cell oxygen cathode which is around 0.6 V. When platinum is used as the anode, the voltage and power produced by a methanol fuel cell will be small, but possibly making them suitable for low power applications.

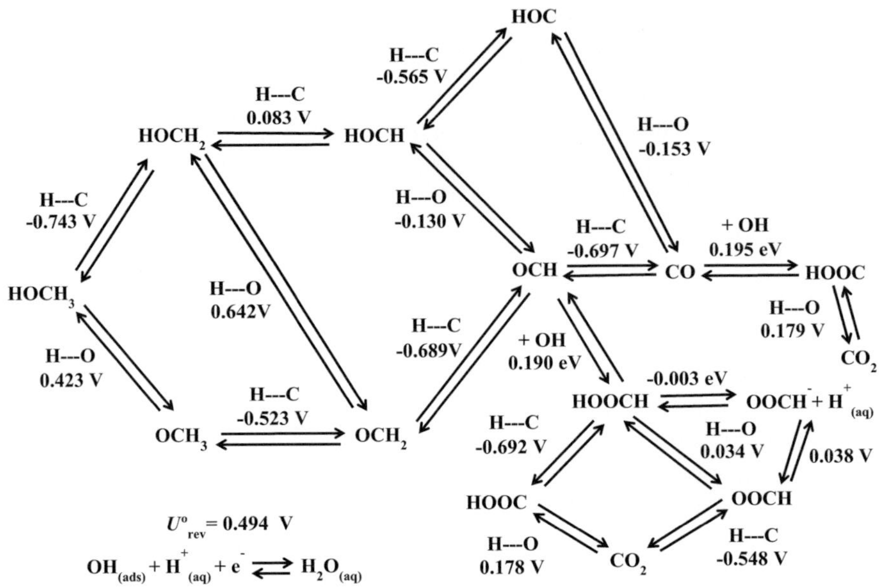


**Fig. 8.16** Reactions and their calculated reversible potentials (V) and, when there was no electron transfer, reaction energies (eV). For clarity the $H^+ + e^-$ that accompany the oxidation products are not shown. All species are adsorbed on Pt(111) except $OCHO^-$(aq) and $H^+$(aq) formed by dissociation of adsorbed formic acid. Reproduced from Ref. [39] with permission from the Royal Society of Chemistry

Trying alternatives to platinum anodes is drawing a lot of attention and the field has been reviewed recently [41].

There was an extensive experimental literature and some theoretical literature concerning formaldehyde and formic acid intermediates at the time the Anderson–Asiri paper was written. For a discussion of this literature, the paper may be consulted.

Based on this LGER study, adsorption energies for intermediates on the ideal methanol oxidation catalyst were calculated. The procedure is to go step by step through the reductions beginning with $CH_3OH$ and ending with $CO_2 + 2H_2O$ and assigning the adsorption energy for the next oxidation intermediate so that the reversible potential for forming it is −0.095 V. This was done for all the pathways in Fig. 8.16 where the ideal adsorption energies, $E^{\mathrm{ideal}}$(ads) and the differences between them and the calculated adsorption energies, $\Delta_{\mathrm{ads}} E$ are in Table 8.5. The most important entry is the 0.590 increase needed for forming OH(ads) from water. This would cause OH(ads) formation from $H_2O$ at −0.095 V making it available for forming HOOC(ads) from CO(ads) which is oxidized to $CO_2 + H_2O$ at a low 0.179 V potential. In fact, with no alterations of the initial adsorption energies, there is a path in Fig. 8.16 where this will be the limiting potential. More electropositive atoms in the surface might form OH bonded to them at lower potentials. But then the energy for OH(ads) adding to CO(ads) might increase from the calculated 0.195 eV in the figure. If CO adsorbed more weakly it would help because it bonds 1.107

**Table 8.5** Ideal calculated adsorption bond strengths at 1/6 ML coverage on Pt(111) based on the calculated bulk solution reversible potentials and $\Delta_{ads} E$ from Table 8.4. $\Delta E$(eV) is the change from $\Delta_{ads} E$. Reproduced from Ref. [39] with permission from the Royal Society of Chemistry

| Molecule | $E^{ideal}$(ads) | $\Delta E$ |
|---|---|---|
| HOOC | 1.998 | −0.322 |
| CO | 0.533 | −1.107 |
| OCH | 1.888 | −0.504 |
| HOC | 3.394 | −1.048 |
| HOCH | 2.635 | −0.470 |
| OOCH | 2.517 | 0.405 |
| HOOCH | 0.386 | 0.275 |
| $HOCH_2$ | 1.483 | −0.648 |
| $OCH_2$ | 0.636 | 0.090 |
| $OCH_3$ | 1.957 | 0.518 |
| OH | 2.541 | 0.590 |

eV too strongly on its present site. Finding a catalyst that that adjusts the adsorption energies for all of the oxidation intermediates in the six-electron reaction may not be possible but focusing on the OH(ads) oxidant is the best way to get started.

An LGER study of ethanol electrooxidation was made in the same format as the above methanol work [42]. As for methanol, it was concluded that for this twelve-electron reduction to $CO_2$ on Pt(111), the optimal surface would have several of the intermediates adsorbing more weakly and OH adsorbing more strongly on the surface.

From these studies it seems that fuel cells using alcohols for fuel will be suitable only for specialized applications where low power is sufficient. For high power applications, such as powering automobiles, $H_2$ is an effective fuel.

There is a strong interest in reducing $CO_2$ to hydrocarbons for use as fuels or chemical feed stocks. Such processes will be in competition with fossil hydrocarbon and plant resources for making these products. Remediation of $CO_2$ because of its contribution to global warming holds a minor secondary interest as long as energy resources remain limited. However, the electroreduction of $CO_2$ is a challenge for fundamental scientific research. In the present context, the equilibrium potentials shown for the pathways in Fig. 8.16 also apply for the reverse reactions which are mechanisms for reducing $CO_2$ on Pt(111). Therefore, beginning with $CO_2$ and $H_2O$ and proceeding backwards, the mechanism encounters steps with negative potentials that require the electrode be set at negative potentials. This will not be possible on the platinum electrode in aqueous electrolyte because $H_2$ is rigorously evolved at negative potentials. On A catalyst with the ideal adsorption energies all reversible potential energies will have the same value whether the goal is oxidizing methanol or reducing carbon dioxide. In either case, the catalyst will need to adsorb many of the intermediates less strongly while it adsorbs OH more strongly.

Electroreduction of carbon dioxide is possible over copper electrodes. Work up to 2019 was reviewed recently [43] Exploring $CO_2$ reduction over copper and copper alloy electrodes continues to attract many research groups.

## 8.16 Concluding Comments

This section introduced frameworks for exploring electrochemistry on electrode surfaces wherein the system energies are calculated at the pzc using semiempirical or ab initio codes. It must be remembered that the adsorption bond strengths of the intermediates on the electrocatalyst surface at the pzc may be sufficiently different at the actual measured reversible potentials that the predicted values maybe significantly distorted, perhaps altering conclusions derived from them. Before moving on to the details of the evolving advanced codes which aim to overcome this deficiency, some results from recent treatments of hydrogen reactions on electrodes using traditional CHE and advanced theories are presented.

## References

1. Anderson AB, Ray NK (1982) Structures and reactions of $H_3O^+$, $H_2O$, and OH on an Fe Electrode: potential dependence. J Phys Chem 86:488–494
2. Davenport AJ, Oblonsky LJ, Ryan MP, Toney MF (2000) The structure of the passive film that forms on iron in aqueous environments. J Electrochem Soc 147:2162–2173
3. Anderson AB, Debnath NC (1983) Mechanism of dissolution and passivation of iron in an aqueous medium: active and transition ranges. J Am Chem Soc 105:18–22
4. Bösing I (2023) Modeling electrochemical oxide film growth—passive and transpassive behavior of iron electrodes in halide-free solution. NPJ Mater Degrad 7:53
5. Anderson AB, Grantscharova E (1995) Potential dependence of CO(ads) oxidation by OH(ads) on platinum anodes. molecular orbital theory. J Phys Chem 99:9143–9148
6. Katayama Y, Kubota R, Rao RR, Hwang J, Giordano L, Morinaga A, Okanishi T, Muroyama H, Matsui T, Shao-Horn Y, Eguchi K (2021) Direct observation of surface-bound intermediates during methanol oxidation on platinum under alkaline conditions. J Phys Chem C 125:26321–26331
7. O'M Bockris J, Wroblowa H (1964) Electrocatalysis. J Electroanal Chem 7:428–451
8. Anderson AB, Grantscharova E (1995) Catalytic effect of ruthenium in ruthenium-platinum alloys on the electrooxidation of methanol. Molecular orbital theory. J Phys Chem 99:9149–9154
9. Anderson AB, Grantscharova E, Seong S (1996) Systematic theoretical study of alloys of platinum for enhanced methanol fuel cell performance. J Electrochem Soc 143:2075–2082
10. Mukerjee S, Srinivasan S (1993) Enhanced electrocatalysis of oxygen reduction on platinum alloys in proton exchange membrane fuel cells. J Electroanal Chem 357:201–224
11. Joyner RW, Shpiro ES, Johnson P, Tuleuova GJ (1993) Intermetallic Pt-Cr clusters in zeolites as models of bimetallic aromatisation and reforming catalysts. J Catal 141:250–265
12. Ma Y, Balbuena P (2008) Surface properties and dissolution trends of $Pt_3M$ alloys in the presence of adsorbates. J Phys Chem C 112:14520–14528
13. Roques J, Anderson AB, Murthi VS, Mukerjee S (2005) Potential shift for OH(ads) formation on the Pt skin on $Pt_3Co(111)$ electrodes in acid. J Electrochem Soc 152:E193–E199
14. Koper MTM, Schmickler W (1998) A unified model for electron and ion transfer reactions on metal electrodes. In: Lipkowski J, Ross PN (eds) Chapter 7 in Electrocatalysis, pp 291–322. Wiley: New York
15. Koper MTM, Voth GA (1998) A three-dimensional potential energy surface for dissociative adsorption and associative desorption at metal electrodes. J Chem Phys 1991–2001

16. Anderson AB, Kang DB (1998) Quantum chemical approach to redox reactions including potential dependence: application to a model for hydrogen evolution from diamond. J Phys Chem A 102:5993–5996
17. Frisch MJ, Trucks GW, Schlegel HB, Gill PMW, Johnson BG, Robb MA, Cheeseman JR, Keith TA, Petersson GA, Montgomery JA, Raghavachari K, Al-Laham MA, Zakrzewski VG, Ortiz JV, Foresman JB, Cioslowski J, Stefanov BB, Nanayakkara A, Challacombe M, Peng CY, Ayala PY, Chen W, Wong MW, Andres JL, Replogle ES, Gomperts R, Martin RL, Fox DJ, Binkley JS, Defrees DJ, Baker J, Stewart JP, Head-Gordon M, Gonzalez C, Pople JA (1995) Gaussian 94 (Revision C.3). Gaussian, Inc. Pittsburgh, PA
18. Mehandru SP, Anderson AB (1985) Dependence of carbon-carbon and carbon-hydrogen bond activation on d band position: acetylene on platinum (111) and iron (100). An electrochemical model. J Am Chem Soc 107:844–849
19. Anderson AB, Albu TV (1999) Ab Initio determination of reversible potentials and activation energies for outer-sphere oxygen reduction to water and the reverse oxidation reaction. J Am Chem Soc 121:11855–11863
20. Anderson AB, Albu TV (2000) Catalytic effect of platinum on oxygen reduction. An ab initio model including electrode potential dependence. J Electrochem Soc 147:4229–4238
21. Kostadinov LN, Anderson AB (2003) Constrained variation calculations of electron-transfer transition states using the lagrange method. Electrochem Solid State Lett 6:E30–E33
22. Aryanpour M, Rai V, Pitsch H (2006) Convergent iterative constrained variation algorithm for calculation of electron-transfer transition states. J Electrochem Soc 153:E52–E57
23. Damjanovic A, Conway BE, Sepa DB (1989) Energetics of an electrochemical reaction derived from the intersection of Tafel lines determined at different temperatures. Bunsenges Phys Chem 93:510–513
24. Damjanovic A, Sepa DB (1990) An analysis of the pH dependence of enthalpies and gibbs energies of activation for $O_2$ reduction at Pt electrodes in acid solutions. Electrochim Acta 35:1157–1162
25. Anderson AB, Cai Y, Sidik RA, B-Kang D (2005) Advancements in the local reaction center reaction center electron transfer theory and the transition state structure in the first step of oxygen reduction over platinum. J Electroanal Chem 580:17–22
26. Magill J, Bloem J, Ohse RW (1982) The mechanism and kinetics of evaporation from laser irradiated $UO_2$ surfaces. J Chem Phys 76:6227–6242
27. Zhang T, Anderson AB (2007) Oxygen reduction on platinum electrodes in base: theoretical study. Electrochim Acta 53:982–989
28. Schmidt TJ, Stamenkovic V, Ross PN Jr, Markovic NM (2003) Temperature dependent surface electrochemistry on Pt single crystals in alkaline electrolyte Part 3. The oxygen reduction reaction. Phys Chem Chem Phys 5:400–406
29. Anderson AB (2012) Insights into electrocatalysis. Phys Chem Chem Phys 14:1330–1338
30. Roques J, Anderson AB (2004) Theory for the potential shift for $OH_{ads}$ formation on the Pt skin on $Pt_3Cr$ in acid. J Electrochem Soc 151:E85-E91
31. Sidik RA, Anderson AB, Subramanian NP, Kumaraguru SP, Popov BN (2006) $O_2$ reduction on graphite and nitrogen-doped graphite: experiment and theory. J Phys Chem B 110:1787–1793
32. Vayner E, Anderson AB (2007) Theoretical predictions concerning oxygen reduction on nitrided graphite edges and a cobalt center bonded to them. J Phys Chem C 111:9330–9336
33. Sidik RS, Anderson AB (2006) $Co_9S_8$ as a catalyst for electroreduction of $O_2$: quantum chemistry predictions. J Phys Chem B 110:936–941
34. Nörskov JK, Rossmeisl J, Logadottir A, Lindqvist L, Kitchin JR, Bligaard T, Jonsson H (2004) Origin of the overpotential for oxygen reduction at a fuel cell cathode. J Phys Chem B 108:17886–17892
35. Anderson AB, Holby EF (2019) Pathways for $O_2$ electroreduction over substitutional $FeN_4$, $HOFeN_4$, and $OFeN_4$ in graphene bulk sites. Critical evaluation of overpotential predictions using LGER and CHE models. J Phys Chem 123:18398–18409
36. Anderson AB (2021) Perspective on theories for reversible potentials for reactions taking place on electrode surfaces. J Electroanal Chem 898:115623

37. Abidi N, Lim KRG, Seh ZW, Steinmann SN (2021) Atomistic modeling of electrocatalysis: are we there yet?" WIREs Comput Mol Sci 11:e1499. https://doi.org/10.1002/wcms.1499
38. Anderson AB, Holby EF (2019) Pathways for $O_2$ electroreduction over substitutional $FeN_4$, $HOFeN_4$, and $OFeN_4$ in graphene bulk sites: critical evaluation of overpotential predictions using LGER and CHE models. J Phys Chem C 123:18398–18409
39. Anderson AB, Asiri HA (2014) Reversible potentials for steps in methanol oxidation and formic acid oxidation to $CO_2$; adsorption energies of intermediates on the ideal electrocatalyst for methanol oxidation and $CO_2$ reduction. Phys Chem Chem Phys 16:10587–10599
40. Iwasita T (2002) Electrocatalysis of methanol oxidation. Electrochim Acta 47:3663–3674
41. Wang J, Zhang B, Guo W, Wang L, Chen J, Pan H, Sun W (2023) Toward electrocatalytic methanol oxidation reaction: longstanding debates and emerging catalysts. Adv Mater 35(1-20):2211099
42. Asiri HA, Anderson AB (2015) Mechanisms for ethanol electrooxidation on Pt(111) and adsorption bond strengths defining the ideal catalyst. J Electrochem Soc 162:F115–F122
43. Nitopi S, Bertheussen E, Scott SB, Liu X, Engstfeld AK, Horch S, Seger B, Stephens IEL, Chan K, Hahn C, Nørskov JK, Jaramillo TF, Chorkendorff I (2019) Progress and perspectives of electrochemical $CO_2$ reduction on copper in aqueous electrolyte. Chem Rev 119(12):7610–7672

# Chapter 9
# Advanced First Principles Methods

**Abstract** A recap of previous chapters is followed by recent advancements. Density functional theory (DFT) is outlined and when implemented in the Vienna *ab initio* simulation package (VASP) it becomes a valuable tool for surface science. Initial applications to electrochemistry explored the effects on surface reaction reversible potentials and activation energies of double layer fields. Fields were introduced by charging the surface and modeling compensation with various forms of background charge. Codes such as "interface" were developed in which the surface is charged and modified Poisson–Boltzmann theory determines the double layer charge distribution variationally. The Fermi level determines the potential. This grand-canonical density functional theory (GC-DFT) approach is the current state of the art and is capable of accurately predicting reversible potentials though obtaining activation energies for electron transfer is computationally demanding. Research is ongoing to improve GC-DFT methods and combine them with molecular mechanics. Mention is made of evolving alternative concepts such as orbital-free DFT and programmable electrodes.

## 9.1 General Introduction

Gurney's picture of resonant electron tunneling to or from the Fermi level to the redox center provided a quantum mechanical basis for the Butler–Volmer description of electron transfer kinetics introduced in the first chapter. For decades Tafel plots and Tafel slopes have been tools for interpreting reaction rates when electron transfers are involved. However, in these models the transferred electron comes from or goes to an energy level with energy determined by the assigned electrode potential but electrode charging and interface polarization are absent, meaning the calculations were for the potential of zero charge (pzc).

Chapters 2 and 3 showed how Marcus developed a model incorporating dielectric properties of the solvent and, through the work of George and Griffiths, thermal vibrations of the ions, to explain potential dependence of electron transfer activation energies. Hush included a double layer potential drop in his model which was based on adiabatic electron transfer and assumed there would be fractional electron charge

 
A. Anderson, *Past and Current Models in Electrochemistry*,
https://doi.org/10.1007/978-3-031-97791-6_9

at the transition state for an electron transfer reaction. Both the Marcus and Hush models omitted the effects of electrode surface charging on the adsorption of reactant molecules and on bond and electrolyte polarizations. Chapter 4 outlined more recent developments using these models.

The Gurney metal band picture was the basis of early quantum chemical approaches introduced in earlier chapters that were aimed at including the effects on adsorbate vibrations caused by changing an electrode's potential. Using shifts in metal surface band energies by means of changing the atomic valence state ionization potential parameters, the semiempirical atom superposition and electron delocalization molecular orbital (ASED-MO) theory yielded sensible predictions of vibrational frequencies and their dependence on applied potential. Though surface-adsorbate bond polarizations varied as the potential was changed, still absent in the model were explicit surface charging and electrolyte response to charging. It was shown how solutions to single determinant electron wave functions in the *ab initio* Hartree–Fock (HF) theory produced energy components which ASED-MO theory modeled. However, HF theory lacked a means to parametrically shift the orbital energy levels to follow changes in applied electrode potential, something ASED-MO theory could do by modifying the experimental atomic orbital valence state ionization potentials used as data in the model Hamiltonian. In Chap. 5 the calculated electrochemical vibrational tuning rate (EVTR) trends were explained using perturbation molecular orbital theory formulas and pictures, originally the basis for understanding bonding and spectroscopy in molecular, particularly organometallic, chemistry. An alternative computational procedure used model electric fields following the definition of Stark tuning rates in gas and vacuum measurements, though the parameterization was uncertain. The ASED-MO band shift technique was also used for rudimentary studies of potential-dependent surface chemistry.

In the 1980s spectroscopies applicable to the study of surface properties and surface chemistry were exploited by electrochemists, allowing them to achieve a greater understanding of adsorbate properties on surfaces under potential control. Some of the widely used spectroscopies are outlined in Chap. 5. Chapter 6 gives an electronic structure-based description of the hydrogen fuel cell. The platinum anode and cathode anode prior to connecting their compartments with a wire and salt bridge were shown to establish double layers with electron chemical potentials −4.4 eV at the anode and −5.6 eV at the cathode. Platinum, with a work function of about 6.0 eV, accommodated this by band bending between the bulk and surface. With the two cells connected by a platinum wire and salt bridge, the electrons enter the wire at the −4.4 eV surface Fermi level ($E_f$) and continue toward the cathode with this chemical potential. Since the work function of the wire surface is 6.0 eV, to describe the electronic structure of the working fuel cell the vacuum level of the wire is shifted up 1.6 eV, corresponding to an adiabatic measurement of workfunction. Along the way to the anode 1.2 eV work is done. Chapter 7 introduced the a relatively new idea of surface conductivity on insulators where conducting states are holes ($h^+$) by electron transfer to acceptor molecules in solution to form a double layer. Optical properties the surface region of insulators with band gap states could be changed by varying the electron chemical potential of an electrolyte because this could change the occupations of the states. Accurate first principle quantum theory had yet to catch up with such measurements, largely because sufficient computing power was unavailable.

Chapter 8 treated catalytic reactions including $O_2$ reduction and CO oxidation as studied using semiempirical molecular orbital calculations and density functional (DFT) calculations. The chapter covered other ground as well. The linear Gibbs energy relationship (LGER) method was introduced. It uses known standard reversible potentials for reactions in bulk solution as starting parameters and perturbed them with calculated reactant and product adsorption energies, yielding predicted reversible potentials for reactions on electrode surfaces. Also proposed was the computational hydrogen electrode (CHE) method which relied on calculated reaction energies on the electrode surface with the proton energy determined using the equilibrium of the standard hydrogen electrode and the electron energy assigned to make the reaction energy zero, corresponding to equilibrium. The electron potential was determined from the difference between its energy and the workfunction of the standard hydrogen electrode (SHE). Studies with these methods continue into the 2020s but still the electrode potential is introduced solely as the electron energy corresponding to the desired potential on the SHE scale, and no surface charging and electrolyte response is included. Studies with these methods often employed modern high accuracy quantum computational codes but new theoretical models were needed to advance understanding and prediction of potential dependencies reactions on electrode surfaces. Improving the accuracy of the DFT codes at the pzc was important but was not a substitute for a full theory including the effects of surface charging accompanying changes in applied potential.

Features of the above-mentioned work included the Arrhenius equation from thermochemistry, the Butler–Volmer equation from the Gurney model, the Dewar-Chatt-Duncanson and other bonding concepts from molecular orbital theory. Understanding based on these models aligned well with the viewpoint of chemists but numerical predictions using them carried a baggage of uncertainty. The developing new spectroscopies opened the door to measurements of greater accuracy and detail, allowing studies in electrochemistry to take on new complexity and subtlety. Understanding and predicting at this new level required quantum mechanical calculations more advanced than the level achieved using simple molecular orbital theory and required kinetic treatments of electron transfer beyond Tafel plots, and the Gurney and Butler–Volmer equations. This need coincided with the development of density functional theory which was implemented by means of parameterized computational codes to give "first principle" predictions, the label applied primarily because the calculations included approximations to electron exchange and correlation energies, distinguishing them from parameterized semiempirical codes such as extended Hückle (EH). The developers and early practitioners of DFT codes were mostly physicists and engineers who had reductionist viewpoint compared to that of many chemists, meaning they had less interest in considering chemical trends and often focused on predicting with highest possible accuracy all measurable properties of an individual system. DFT was an entry into this world while molecular orbital theory was not. A third computational genre, *ab initio* theory with large basis sets and multiple electronic configuration (CI) wavefunctions provided the most accurate predictions but was effective only for small molecules and so has not found many applications to electrochemistry. While accurate predictions with CI are possible for small molecules, finding chemical interpretations of the kind that are made when using MO theory can be complicated and problematic because of the multiple configuration basis functions in the calculations. This causes practitioners to focus on calculated energies and structures. DFT as applied with approximations gives orbital wavefunctions and energies that are compatible with molecular orbital interpretations of structures and energy changes, making them fit the established language of chemists. Most importantly, DFT calculations were not limited in application to small molecules but could be performed for large clusters and extended surfaces via band theory.

The local reaction center (LRC) model had its genesis in 1998 and was applied to steps in oxygen reduction and hydrogen oxidation on platinum using *ab initio* and hybrid density functional theories. The approach provided suggestive predictions of potential-dependent electron transfer activation energies and reversible potentials for surface reactions involving electron transfer. However, the necessity of using one or two atoms to represent the electrode surface meant it could not be improved through the use of more realistic surface models. Consequently, after a few years its usage declined. Moreover, LRC calculations were performed at the pzc and a handle on the consequences of this constraint was not yet at hand. What was needed for predicting potential-dependent activation energies was a band theory approach that would be usable for modeling adsorption, chemical reactions, and electron transfer on extended surfaces. Whereas semiempirical band theory codes existed and were used primarily for structural studies, for accurate electrochemical work electron repulsions had to be accurately included in the calculations. This became possible with the advent of density functional theory which was used for the LRC, CHE, and LGER calculations discussed in Chap. 8. Chapters 10 and 11 include applications of DFT band theory calculations to hydrogen adsorption and oxygen reduction. At this point it is appropriate to outline some of the details of DFT.

## 9.2 Summary of Density Functional Theory (DFT)

The density functional computational models and codes developed in the 1990s are the practical but approximate outgrowths of the density functional theory created by Hohenberg and Kohn [1] and Kohn and Sham [2]. The density functional equations, discovered Hohenberg and Kohn, gave the energy of a system as a functional of the electron charge density distribution function. As in HF theory, the electronic energy of a molecule in a given structure is the sum of kinetic electron kinetic energy, $T$, nuclear-electron attraction, $V_{\text{ne}}$, and electron repulsion energies, $V_{\text{ee}}$, but in density functional theory each depends on the charge density distribution function, $\rho$:

$$E[\rho] = T[\rho] + V_{\text{ne}}[\rho] + V_{\text{ee}}[\rho] \tag{9.1}$$

For a given structure $V_{ne}$ is found by integration but the $T[\rho]$ and $V_{ee}[\rho]$ components are not and are combined in the Hohenberg–Kohn universal functional $F_{HK}$:

$$E[\rho] = \int \rho(r)v(r)d(r) + F_{HK}(\rho) \tag{9.2}$$

In Eq. (9.2) the operator in the integral is the nuclear-electron attraction operator which gives rise to $V_{ne}[\rho]$. Unfortunately, the functional was unknown and likely would never be known. Despite this, the work of Hohenberg and Kohn were inspired to look beyond the free electron gas theory, for which the energy was known for systems of constant electron density. Kohn and Sham introduced electron orbitals in determinants as employed in Hartree–Fock theory. The energy was partitioned into coulomb, exchange correlation, and kinetic components; these terms being understood from Hartree–Fock single determinant wavefunction quantum mechanical calculations for molecules and solids. The total energy took the form

$$E = \sum \varepsilon_i - 1/2 \int \rho(r)\rho(r')|r - r'|^{-1} d(r)d(r') + E_{xc}(\rho) - \int v_{xc}(r)d(r) \tag{9.3}$$

$E_{xc}$ is exchange correlation energy and $v_{xc}(\boldsymbol{r})$ is exchange correlation potential. The sum is over N electrons and is

$$\begin{aligned}\sum \varepsilon_i &= \sum \langle \Psi_i(r)| - 1/2\nabla^2(r) + v_{eff}(r)|\Psi_i(r)\rangle \\ &= T[\rho] + \int v_{eff}(r)\rho(r)d(r)\end{aligned} \tag{9.4}$$

In Eq. (9.4) $v_{eff}(\boldsymbol{r})$ is the Kohn–Sham effective potential, given by

$$v_{eff}(r) = v(r) + \int \rho(r')|r - r'|^{-1} d(r') + v_{xc}(r) \tag{9.5}$$

As in Hartree–Fock theory, the electron energy is the sum of orbital energies with components of the total electron repulsion energy subtracted from it. Finding the exchange correlation energy and the exchange correlation potential is the challenge in applying density functional theory to predicting energies of molecules and solids. In practice, the exchange correlation components of the energy formula may be seen as elaborations on the formula for the energy of a free electron gas. Formulas using a local density approximation (LDA) were developed and additional parameterized terms incorporating density gradients were added for determining approximate values for the exchange correlation energy and potential. With parameters chosen to make the calculations reproduce selected experimental data, the theory becomes essentially semiempirical. It is quite accurate and generally useful semiempirical theory, but to avoid confusion with existing semiempirical codes, the description "first principles" became a commonly used adjective connected to DFT.

The orbital energies, $\varepsilon_i$, in DFT have a very useful property insofar as applications to electrochemistry are concerned. By the Janak theorem, $\varepsilon_i$ in DFT is interpreted the same way as Koopman's ionization potential in HF theory [3]. Helpfully, the density functional equations also apply when fractional electrons are treated, that is, when $N$ is not an integer. For molecules with low densities of states, meaning significant gaps between the orbital energies, ionization energies and electron affinities are approximately averages of the HOMO or LUMO energies for the neutral and charged molecule. For metal electrodes the density of states is high so that both $IP$ and $-EA$ have the value − the Fermi energy.

Important follow-up studies exist for the Hohenberg–Kohn and Kohn–Sham papers along with many extensions and applications in chemistry and physics. A comprehensive monograph by Parr and Yang gives a presentation with exemplarily clarity, and the above presentation follows sections of it [4]. Properties of solids and surfaces are theoretically understood in terms of band theory and DFT was first incorporated into band theory with plane wave basis sets. The essence of band theory is to define the smallest translational cell that provides by translation the atoms necessary to adequately replicate the bulk or surface, like stacking bricks ad infinitum. The bulk or surface wavefunctions, energies, and density of states are calculated with this basis. A surface process on one brick occurs simultaneously on all the others and so as the bricks are made larger, interactions between the surface species can be reduced but computational costs increase rapidly.

## 9.3 DFT Applications to Electrochemistry

The first readily available and widely used band theory DFT code was the Vienna *ab initio* simulation package (VASP) introduced in a series of papers beginning in 1993 [5–8]. Their initial description as "*ab initio*" was overtaken by "first principles" to avoid confusion with already existing *ab initio* methods that were characterized by their use of the full Hamiltonian for all electrons.

When using plane waves for determining wave functions for a three-dimensional system, the basis includes waves in three Cartesian directions to generate a Fourier series. There are no impediments for infinite three-dimensional crystalline structures. Two-dimensional slab-like systems must be calculated like crystals but with layers of slabs repeating ad infinitum. These may be referred to as DFT slab-band calculations. To reduce inter-slab interactions, the chosen inter-slab distance is usually around 10 Å, or more. Convergence to useful limits in wave functions and energies demands more plane waves of shorter and shorter wavelength in the basis set normal to the planes and therefore need more computational time. One-dimensional systems require more plane waves in both directions normal to the chain and still more computational effort.

An alternative to using plane waves in computations with DFT is to use atomic orbital (AO) basis sets. When this is done there is no need for using repeating separated slabs for two-dimensional systems used to model surface chemistry or repeating separated chains to model chains such as nanowires. This alternative approach was developed for electrochemistry applications more recently and will be introduced later.

Applications of VASP to solid and surface system properties and electronic structures soon became abundant. Surface calculations with plane waves use the repeating two-dimensional slab formalism. About ten years passed before the VASP code was applied to calculating approximate reversible potentials for intermediates formed during reactions on electrode surfaces. Reduction of oxygen to water, the important reaction at platinum fuel cell cathodes, was a common topic. The early electrochemical applications of the VASP code were calculating reversible potential shifts for reactions on platinum alloy surfaces using vacuum interface models. These were followed by calculations of reversible potentials on transition metals by the CHE and LGER methods. The effects of alloying Pt surfaces were treated. Calculations of activation energies for coupled electron, proton transfers through the use of the local reaction center model introduced earlier were also performed. Such work employed open reaction systems, wherein the electron is introduced from or sent to an external reservoir assigned to an energy on the vacuum scale corresponding to the chosen potential. Bonding polarizations between the electrode surface and adsorbed molecules are present but the potential of the electrode itself is not controlled. This means calculations were performed at the potential of zero charge (pzc). The Fermi energies in the self-consistent DFT slab-band calculations vary with the molecules adsorbed and the electrode workfunction depends on surface species but is not tied to an applied potential and is not controlled. As explained earlier, the reaction reversible potential predictions depend on reaction energies (CHE) or adsorption energies of reactants and products (LGER). Applications such as predicting reversible potentials for one-electron steps in multi-electron reactions such as oxygen reduction were made using both models.

## 9.4 DFT Double Layer Modeling and Reaction Energies

In the mid-2000s, researchers added explicit surface charging to calculations. One VASP-based model compared the approach used in LGER and CHE calculations, where the potential is ascribed to the electron energy alone, with two models, one where electric fields were imposed and the other where electrons with compensating diffuse background charge were added [9]. The first and third methods predicted lower reversible potentials for O(ads) formation than for OH(ads), in disagreement with measurement but the authors nevertheless concluded that the first model, which is the simplest to calculate, accounted very well for the potential dependencies of the surface phases. The method of surface charging accompanied by a diffuse background countercharge was used in follow-up studies [10, 11]. Others picked up this approach and ran with it, for example an oxygen reduction study over Pd(111) and Pt(111) [12]. This study based its conclusions on calculated reaction and activation energies but not on reversible potentials for the steps. Conclusions based on these studies lacked certainty because they omitted the key feature of electrochemistry which is the potential applied to the electrode, which was not controlled in the studies so that the actual potential varied unavoidably throughout the course of the reactions.

A method for changing applied potential where a charged plane or Gaussian distribution outside the surface served as counter charge was also being explored [13, 14]. As implemented, this model could maintain a fixed electrode potential during the course of a reaction. More is said on this below. Skúlason and coworkers invoked a model with electrons added to the slab-band translational cell with charge-neutralizing hydronium ions in the double layer [15]. Due to the integral number of hydronium ions in a translational unit cell used in the DFT calculations, only widely spaced potentials could be modeled, and there was no provision for maintaining a given potential during the course of a reaction.

In an improvement over the various background charge models, Otani and Sugino introduced a modified Poisson–Boltzman (MPB) theory for ion distribution into the DFT calculations [16]. Jinnouchi's code for the electrochemical interface used a smooth flexible dielectric function [17]. In place of the plane wave basis, this work followed of Soler et al. and employed linear combinations of pseudoatomic orbitals rather than plane waves [18]. Surfaces were represented by slabs by using translational unit cells in two dimensions. The code included options to calculate one-dimensional chains and three-dimensional solids. The electrode potential keyed directly with $E_f$, which was adjusted by surface charging in this grand-canonical density functional theory, GC-DFT, approach. This code, referred to as "interface," was employed already in Chaps. 5 and 7 and it gave good predictions of electrochemical vibrational tuning rates (EVTR) and adsorption energies at the pzc. It was also used in the LGER study of methanol electrooxidation. Full details, including literature sources and background, are in ref. [17] and there the theory was already applied to the effect of solvation and the potential dependence of surface energies before and after $O_2$ adsorption.

Initial tests of the interface code were calculations of 34 aqueous redox potentials over a six volt range [19]. Comparison with experiment is in Fig. 9.1. Two functionals were used with the generalized gradient approximation, Perdew-Burke-Ernzerhof (PBE) and revised PBE (RPBE). As discussed in the above Phys. Rev. B paper [17], calculations with the PBE functional overestimate PtO bond strengths and calculations with the RPBE functional gave results in better agreement with experiments. However, the absolute errors for the aqueous reactions were 0.19 V for RPBE and 0.13 V for PBE.

Reversible potentials for reactions of adsorbed species may be found using interface by calculating the energies of the reactant and product at different potentials separately and plotting them on the same graph. The potentials are changed by adding or removing approximately 0.1 e increments of electron charge to the translational cell in the slab-band model of the electrode surface. Where the curves cross the reactant and product energies are the same and the corresponding potential is the predicted reversible potential. Figure 9.2 gives an example and additional examples are in later chapters.

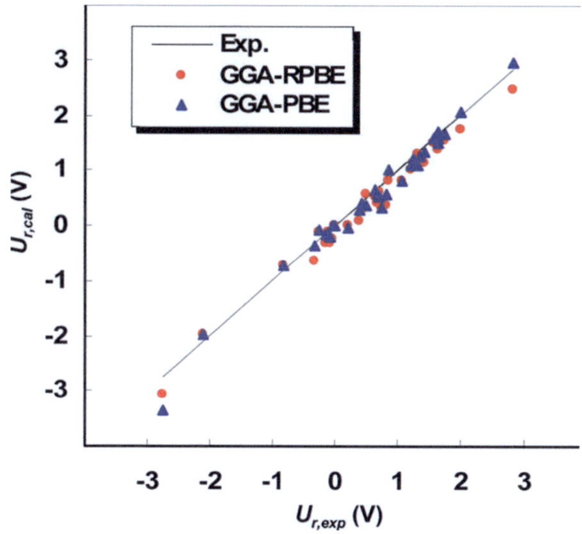


**Fig. 9.1** This graph of calculated reversible potentials vs experimental values for reactions over a nearly 6 V range shows the predictions using RPBE and PBE functionals are about the same but on average slightly underestimate experiment. Reprinted with permission from R. Jinnouchi and A. B. Anderson, "Aqueous and Surface Redox Potentials from Self-Consistently Determined Gibbs Energies," J. Phys. Chem. C 112, 8747–8750, 2008. Copyright 2008 American Chemical Society

Pt surface redox potentials were calculated to various levels of approximation using both functionals. Four reactions relevant to $O_2$ reduction at fuel cell cathodes were treated:

$$H^+(aq) + e^- \rightleftharpoons H(ads) \tag{9.6}$$

$$H^+(aq) + O_2(g) + e^- \rightleftharpoons HO_2(ads) \tag{9.7}$$

$$H^+(aq) + OH(ads) + e^- \rightleftharpoons H_2O(ads) \tag{9.8}$$

$$H^+(aq) + O(ads) + e^- \rightleftharpoons OH(ads) \tag{9.9}$$

Four levels of calculation were applied to the reactions:

(i) The use of internal energies of adsorbates on neutral surfaces.
(ii) The use of Gibbs free energies of adsorption on neutral surfaces.
(iii) Adding solvation Gibbs free energies from MPB to (ii).
(iv) Adding surface charging to (iii).

The mean absolute errors in the calculated values compared with experiment for the four levels was smallest for (i) and (iv). In the case of RPBE they were for the

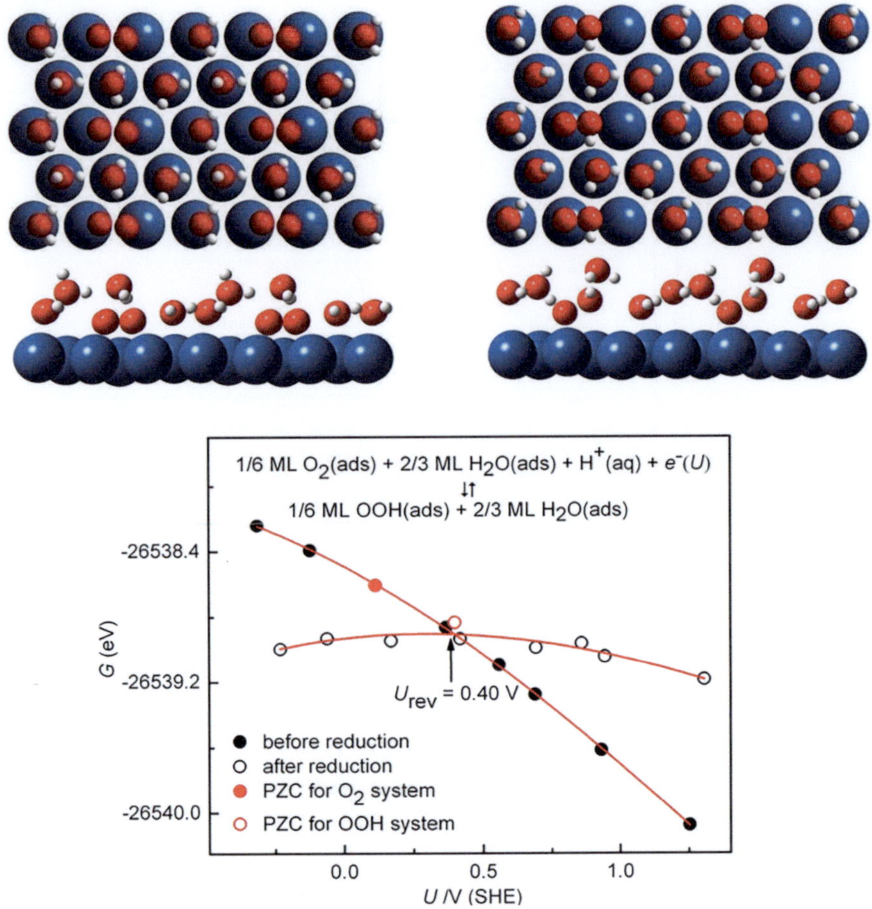


**Fig. 9.2** Views of 1/6 ML $O_2$(ads) + 2/3 ML $H_2O$ at top left and 1/6 ML OOH(ads) 2/3 ML $H_2O$ at top right. Pt atoms-blue, O atoms-red, H atoms-white. The graph shows the energy curves for reactant and product crossing at $U_{rev} = 0.40$ V. Charge increments are approximately 0.1 electron charge units. Calculated pzcs are indicated on the curves. Reprinted with permission from F. Tian and A. B. Anderson, "Effective Reversible Potentials, Energy Loss, and Overpotential on Platinum Fuel Cell Cathodes," J. Phys. Chem. C 115, 7076–7088, 2011. Copyright 2011 American Chemical Society

four levels 0.18, 0.34, 0.24 and 0.23 V, while for PBE they were 0.19, 0.25, 0.21, and 0.19 V. Results with both functionals predicted lower reversible potentials for Eq. (9.9) than for Eq. (9.8), which is incorrect. With the addition of one adsorbed $H_2O$ molecule with OH(ads), generating stabilizing hydrogen bonds between OH(ads) and $H_2O$(ads), the predictions took on the correct order and the mean error for level (iv) RPBE calculations dropped from 0.25 V to 0.13 V and raised from 0.21 V to 0.29 V for PBE calculations.

As the author summarized in 2017, an advanced theory for the electrochemical interface has the following features: a. charge is added; b. the $E_f$ shifts; c. the electrolyte is polarized; d. bonds are polarized; e. electron transfer may occur [20]. The interface code has these and within a few years other researchers were developing and applying similar codes with these features [21–24]. Work also continued on development of the Poisson–Boltzmann theory [25, 26].

The use of MPB theory for calculating solvation energies provides usefully accurate predictions for low ion concentrations and ongoing development by various labs was recently reviewed [27]. Jinnouchi's overview of GC-DFT shows where and how previous theories fit within the grand-canonical picture [28].

## 9.5 DFT Approaches to Calculating Transition States and Activation Energies

Filhol and Neurock introduced a model for the water-electrode interface that provided for changes in applied potential [29]. In their model, electrons were added to or subtracted from the translational cell for the DFT slab with resulting total charge $q$. An hexagonal bilayer of water molecules representing part of the double layer was added to the translational cell and the whole translational cell was bathed in a uniform charge distribution to maintain its neutrality. Taking into account the resulting average electrostatic potential the authors come to the formula

$$E_{\text{free}} = E_{\text{dft}} + \int \langle V_{\text{tot}}(Q) \rangle dQ - qE_{\text{f}} \tag{9.10}$$

Here $E_{\text{free}}$ is the Helmholtz free energy, the integral is from 0 to $q$, $< V_{\text{tot}}(Q) >$ is the average potential in the unit cell, and $E_f$ is referenced to an internal vacuum. Water reactions were calculated over Pd(111). The energies of three different states of water, $H_2O$(ads), OH(ads)/$H_3O^+$, and H(ads)/$OH^-$ were calculated and graphed as functions of charge. For the reaction in acid forming OH(ads), the $H_2O(\text{ads}) \rightleftharpoons \text{OH(ads)} + H_3O^+(\text{aq}) + e^-$ the calculated equilibrium potential was 1.1 V (SHE). Though 0.2 V to 0.4 V greater than quoted experimental values, this was considered good agreement at the time. For the reaction in base forming H(ads), the $H_2O(\text{ads}) + e^- \rightleftharpoons \text{H(ads)} + OH^-(\text{aq})$ equilibrium potential was calculated to be 0.5 V (SHE), which is close to a quoted approximate experimental value of 0.4 V. However, the experimental value for the reaction in base and was reported on the RHE scale, not the SHE scale. Too place it on the SHE scale would mean subtracting 0.83 V, and this removes the apparent agreement. To predict the reversible potential for forming H(ads) on the SHE scale the reaction to calculate is $H_3O^+(\text{aq}) + e^- \rightleftharpoons \text{H(ads)} + H_2O$. This brings up the sometimes confusing aspects of pH changes and the relationship between RHE and SHE. More on pH will follow in a later section.

In a later paper the Neurock group extended the model to calculating transition states and activation energies for reducing $O_2$ adsorbed on Pt(111) to OOH(ads)

[30]. Zero point vibrational energies and vibrational entropy contributions $TS_{\text{vib}}$ were added to the initial and final state Helmholtz energies to approximate the reactant and product Gibbs energies. A reversible potential of 0.5 V was calculated which is close to 0.34 V calculated by Jinnouchi and Anderson with the RPBE functional. The experimental value was not known. Using a nudged elastic band method, the authors determined transition states at constant charges, obtaining activation energies as Helmholtz free energies of activation. They calculated an activation energy of 0.31 eV at 0.8 V which was not much larger than the quoted 0.26 eV experimental value [31]. Energy profiles and activation energies calculated at constant charge by the nudged elastic band method will not generally be the activation energies calculated or measured at constant potential. At constant charge $E_{\text{f}}$, and therefore electrode potential, is sensitive to structure and both continually change along the reaction path. In experiments the system is open so to maintain the potential electronic charge flows between the system and its surroundings. DFT calculations at fixed charge give Helmholtz energies and to maintain constant $E_{\text{f}}$ the calculation model must be open to electron charge transfer in or out, and in DFT these can be fractional electron charges. Details for doing this are available in the work of Lozovoi et al [13].

Jinnouchi with his coworkers at Toyota Central R&D labs employed the interface code to calculate activation Gibbs energies for reactions (9.8) and (9.9) in forward and reverse directions over the 0.6–1.2 V Pt(111) electrode potential range [32]. Their work yielded predictions of cyclic voltammograms and Tafel slopes in good agreement with the measured irreversibility of the former and small values of the latter. Electron delocalization in the transition state was seen, in agreement with the Hush model. The work required many self-consistent calculations to determine reaction coordinates and energies for each potential point.

Soon after the Jinnouchi study, Head–Gordon and coworkers published a VASP-based code including surface charging, a Poisson–Boltzmann model for electrolyte polarization, and a continuum dielectric model for the solvent to calculate transition state structures and activation Gibbs energies for CO coupling reactions on Cu(111) electrodes [33]. They also, as did Jinnouchi et al., found the work computationally intensive, and they reported requiring up to 10 times the number of calculations at unconstrained potential to locate an energy for a particular potential. The factor reduced to 3 times for subsequent calculations by extrapolating from prior results.

Two faster algorithms were derived by Sundararaman, Goddard and Arias wherein the electrode potential rather than fixed charges are starting points in the calculations [34]. They applied their algorithms to settle a question about interpreting two upd voltammometric peaks for deposition on a Pt(111) electrode from a copper chloride solution. The application included a semiempirical component, use of the experimental standard reduction potential for $Cu^{2+}$(aq) because of the difficulty in accurately calculating its free energy. The higher potential peak was found to be from the desorption of chloride ions and not desorption from a lower coverage copper phase as some suggested. The new algorithms [35] were used in work discussed in Chap. 7, and for other mechanism and reaction rate predictions [36]. For additional perspective, one can see various reviews [37].

## 9.6 Relationships of State Changes in DFT and Electrostatics to the pzc

Since the early days of electrochemistry there has been interest in measuring and predicting the potential of zero charge (pzc). It appears in as a parameter in Gurney's band picture for electron transfer reactions. In their comprehensive book Bockris and Kahn viewed it as a fundamental idea of electrochemistry [38]. It has been a continuing focus since then. Experimentally, the pzc is the potential where the double layer capacitance is at a minimum. As will be shown in Chap. 12, changes in overlayer $H_2O$ structure may take place as the potential passes through the pzc but the relationships, if any, of the pzc to electrocatalysis remain to be demonstrated and a general relation may not exist.

In a DFT 1- or 2- or 3-dimensional band calculation on a metal, the highest occupied orbital has energy identified as $E_f$, and is expressed relative to zero for the vacuum. Experimentally, an electron removed from or added to the Fermi level comes from or goes to a state delocalized over the whole infinite system so it adds zero to the surface charge density. Therefor the process causes a nil perturbation on the band electronic structure and $E_f$ is the ionization potential or negative electron affinity of the surface. When the surface is perturbed by immersion into an electrolyte, $E_f$ in most cases is shifted to a new value due to bonding interactions such as $H_2O$ lone-pair donation to surface atoms but $E_f$ still represents the *IP* or $-EA$ for putting an electron into the vacuum level or adding an electron from the vacuum level. When no charge density is added to the surface, the surface potential remains at the pzc. Potential may be reported relative to the SHE or on some other scale such as RHE or SCE. If electronic charge is taken from or added to the surface of this "working" electrode by a metal wire and electrolyte connected to counter electrode, the ions in the electrolyte will adjust so that the net charge in the interfacial region is zero. The driving force is the Gibbs energy difference between the two half-cell reactions. The counter electrode will also bear electronic charge and have a compensating distribution of interfacial ions of net opposite charge to the working electrode. The charge on the working electrode changes $E_f$ and again the *IP* or $-EA$ depend just on $E_f$. For more detail, see Chap. 6.

The previously discussed work of Hansen et al. and Kötz et al. showed following emersion from the electrolyte the workfunctions of electrodes initially under potential control changed linearly with changes in the applied potential. In these measurements the emitted electrons apparently passed elastically through the attached compensating double layers which should have experienced partial dehydration. Structure variations in the double layers probably accounted for the observed data point scatter but the linear trends were clear.

DFT allows modeling the potential dependence of an electrochemical interface due to surface charging when a polarizable electrolyte is part of the computational model. The potential of the electrode will be $[E_f(\text{SHE}) - E_f(\text{electrode})]/F$ where $E_f(\text{SHE})$ is approximately $-4.43$ eV.

The pzc is the easiest potential to model theoretically since no added surface electronic charge and compensating double layer are part of the calculations. There is a history of development of electrostatic approaches to calculating the pzc from interface models with pure water overlayers. Santos and Schmickler give a recent overview [39]. In the electrostatic approach the electrostatic profile is calculated from the bulk of the metal through interface where water molecules may be adsorbed, through a region of bulk water and then through the water surface to vacuum. The authors note errors of several volts with this approach. They proposed calculating the pzc when using DFT as the potential where the derivative of the surface energy with charge is zero. This was already seen in a DFT study of Tian and Anderson in 2011 [40].

DFT calculations of the electrochemical interface that use surface charging and compensating charge response in the electrolyte provide predictions for potential dependence of molecules adsorbed or in the double layer region. These predictions obviate the need to address electric fields, though they may be calculated using the self-consistent charge distribution functions. This is a simple and direct approach to calculation the effects of potential changes in electrochemistry.

## 9.7 Programmable Electrodes

A new dimension is currently being added to the field of heterogeneous catalysis of chemical synthesis. It is materials programing and its development is reviewed by Dauenhauer, who also is a major contributor [41]. The idea is to expose the catalyst to energy to change the bond strength of reaction intermediates on the active sites and this can include radiative energy to change electronic states or forces applied to stress the material and change the atom positions. The hope and goal is to find a temporal sequence of such perturbations that maximizes the efficiencies of a multistep reaction over known and potential catalyst materials. In the electrochemical domain an example of the idea is clearing platinum electrode of CO(ads) and other carbon impurities by oxidative cycling to increasing potentials or removing OH(ads) by reductive cycling to decreasing potentials.

Thinking about the four-electron reduction of $O_2$ on a cathode to $H_2O$ in a fuel cell, it is at this stage not possible, at least for the author, to imagine a material and the temporal energy input manipulations that might make an electrode made of this material operate at maximum potential. Further discussion of the constraints is in Chap. 11. Keeping in mind that a fuel cell is used for producing electrical energy, the important question is: would it provide enough energy to execute those manipulations while still providing useful output?

The exploration of programmable heterogeneous catalysts for chemical synthesis is a relatively new development. It is an endeavor to keep an eye on. Making products is the goal, and successful processes will not use prohibitive quantities of energy. Developing programmable catalysts for electrochemical synthesis is a worthwhile and exciting research goal and could be revolutionary. However, energy generation using programmable electrocatalysts has a conflict of interest between energy input and output, a conflict that is not present for chemical synthesis. It is hard to imagine that efficient programmable electrodes can be developed for oxygen reduction in fuel cells.

## 9.8 Evolving Methods

Orbitals are introduced to Kohn–Sham density functional theory to give a method for calculating electron kinetic energies. An alternative is orbital-free DFT (OFDFT) wherein functionals are devised for determining the electron kinetic energies. Computations are faster in application than for KS-DFT and they allow for constant potential description of electrochemical double layers [42]. Huang calls the theory density-potential functional theory (DPFT). A subsequent study showed how, by using a parameterized description of the electrolyte molecules, a detailed description of the electrolyte structure approaching an electrode surface could be achieved [43]. It will be interesting to see future studies with OFDFT and DPFT.

Steinmann has found that for polar sites on $MoS_2$ quantum mechanical molecular mechanics (QM/MM) methods may be more accurate than GC-DFT codes like interface [44]. Such computationally intensive codes may be necessary when calculating molecular adsorption energies on polar sites but the study also found that QM/MM calculated potential-dependent Gibbs energies for non-polar sites were close to those calculated with a VASP GC-DFT code. Jinnouchi published a theoretically and mathematically rigorous introduction to the evolving advanced theories including interface and related codes and it includes a good survey of the literature pertaining to molecular dynamics (MD) and machine learning (ML) potentials [28]. Interested readers are directed to this paper. An upcoming review by Jinnouchui and Minami is an outstanding survey of artificial intelligence (AI) and ML covering their advancing development and applications to materials and electrochemistry [45]. For a phenomenological approach to explaining theoretical treatments of electrode potential a recent review by Anderson may be consulted [46].

## References


1. Hohenberg P, Kohn W (1964) Inhomogeneous electron gas. Phys Rev 136:B864–B871
2. Kohn W, Sham LJ (1965) Self-consistent the Hohenberg-Kohn-equations including exchange and correlation effects. Phys Rev 140:A1133–A1138
3. Janak JF (1978) Proof that $\partial E/\partial n_i = \varepsilon_i$ in density-functional theory. Phys Rev 18:7165–7168
4. Parr RG, Yang W (1989) Density-functional theory of atoms and molecules. Oxford UP, New York
5. Kresse G, Hafner J (1993) Ab initio molecular dynamics for liquid metals. Phys Rev B 47:558
6. Kresse G, Hafner J (1994) Ab initio molecular-dynamics simulation of the liquid-metal-amorphous-semiconductor transition in germanium. Phys Rev B 49:14251

7. Kresse G, Furthmüller J (1996) Efficiency of ab-initio total energy calculations for metals and semiconductors using a plane-wave basis set. Comput. Mat. Sci. 6:15
8. Kresse G, Furthmüller J (1996) Efficient iterative schemes for ab initio total-energy calculations using a plane-wave basis set. Phys Rev B 54:11169
9. Rossmeisl J, Nørskov JK, Taylor CD, Janik MJ, Neurock M (2006) Calculated phase diagrams for the electrochemical oxidation and reduction of water over Pt(111). J Phys Chem B 110:21833–21839
10. Taylor CD, Wasileski SA, Filhol J-S, Neurock M (2006) First principles reaction modeling of the electrochemical interface: consideration and calculation of a tunable surface potential from atomic and electronic structure. Phys Rev B 73(16):165402
11. Janik MJ, Taylor CD, Neurock M (2007) First principles analysis of the electrocatalytic oxidation of methanol and carbon monoxide. Top Catal 46:306–319
12. Ou L, Chen S (2013) Comparative study of oxygen reduction reaction mechanisms on the Pd(111) and Pt(111) surfaces in acid medium by DFT. J Phys Chem C 117:1342–1349
13. Lozovoi AY, Alavi A, Kohanoff J, Lynden-Bell RM (2001) Ab initio simulation of charged slabs at constant chemical potential. J Chem Phys 115:1661–1669
14. Lozovoi AY, Alavi A (2003) Reconstruction of charged surfaces: general trends and a case study of Pt(110) and Au(110). Phys Rev B 68(18):245416
15. Skúlason E, Karlberg GS, Rossmeisl J, Bligaard T, Greeley J, Jónsson H, Nørskov JK (2007) Density functional theory calculations for the hydrogen evolution reaction in an electrochemical double layer on the Pt(111) electrode. Phys Chem Chem Phys 2007(9):3241–3250
16. Otani M, Sugino O (2006) First-principles calculations of charged surfaces and interfaces: a plane-wave nonrepeated slab approach. Phys Rev B 73(11):115407
17. Jinnouchi R, Anderson AB (2008) Electronic structure calculations of liquid-solid interfaces: combination of density functional theory and modified Poisson-Boltzmann theory. Phys Rev B 77(18):245417
18. Soler JM, Artacho E, Gale JD, García A, Junquera J, Ordejón P, Sánchez-Portal D (2002) The Siesta method for ab initio order-N materials simulation. J Phys Condens Matter 14:2745–2779
19. Jinnouchi R, Anderson AB (2008) Aqueous and surface redox potentials from self-consistently determined Gibbs energies. J Phys Chem C 112:8747–8750
20. Anderson AB (2017) Concepts and computational methods for the electrochemical interface and applications: past, present and future. Curr Opin Electrochem 1:27–33
21. Weaver KL-, Arias TA (2012) Joint density functional theory of the electrode-electrolyte interface: application to fixed electrode potentials, interfacial capacitances, and potentials of zero charge. Phys Rev B 86(16):075140
22. Fang YH, Wei GF, Liu ZP (2013) Theoretical modeling of electrode/electrolyte interface from first-principles periodic continuum solvation model. Catal Today 202:98–104
23. Mathew K, Sundararaman R, Weaver KL-, Arias TA, Hennig RG (2014) Implicit solvation model for density-functional study of nanocrystal surfaces and reaction pathways. J Chem Phys 140(8):084106
24. Xiao H, Cheng T, Goddard WA III, Sundararaman R (2015) Mechanistic explanation of the pH dependence and onset potentials for hydrocarbon products from electrochemical reduction of CO on Cu(111). J Am Chem Soc 138:483–496
25. Fisicaro G, Genovese L, Andreuss O, Mazari N, Goedecker S (2016) A generalized Poisson and Poisson-Boltzmann solver for electrostatic environments. J Chem Phys 144(12):014103
26. Ringe S, Oberhofer H, Hille C, Matera S, Reuter K (2016) Function-space-based solution scheme for the size-modified Poisson-Boltzmann equation in full-potential DFT. J Chem Theory Comput 12:4052–4066
27. Ringe S, Hörmann NG, Oberhofer H, Reuter K (2022) Implicit solvation methods for catalysis at electrified interfaces. Chem Rev 122:10777–10820. https://doi.org/10.1021/acs.chemrev.1c00675
28. Jinnouchi R (2024) Grand-Canonical first principles-based calculations of electrochemical reactions. J Electrochem Soc 171:096502

29. Filhol JS, Neurock M (2006) Elucidation of the electrochemical activation of water over Pd by first principles. Angew Chim. 118:416–420
30. Janik MJ, Taylor CD, Neurock M (2009) First-principles analysis of the initial electroreduction steps of oxygen over Pt(111). J Electrochem Soc 156:B125–B135
31. Damnovic A, Sepa DB (1990) An analysis of the pH dependence of enthalpies and Gibbs energies of activation for $O_2$ reduction at Pt electrodes in acid solutions. Electrochim Acta 35:1157–1162
32. Jinnouchi R, Kodama K, Suzuki T, Morimoto Y (2015) Kinetically induced irreversibility in electro-oxidation and reduction of Pt surface. J Chem Phys 142(184709):1–13
33. Goodpaster JD, Bell AT, Head-Gordon M (2016) Identification of possible pathways for C-C bond formation during electrochemical reduction of $CO_2$: new theoretical insights from an improved electrochemical model. J Phys Chem Lett 7:1471–1477
34. Sundararaman R, Goddard III WA, Arias TA (2017) Grand canonical electronic density-functional theory: algorithms and applications to electrochemistry. J Chem Phys 146(14):114104
35. Liu G, Lee M, Kwon S, Zeng G, Eichhorn J, Buckley AK, Toste FD, Goddard III WA, Toma FM (2021) $CO_2$ reduction on pure Cu produces only $H_2$ after subsurface O is depleted: theory and experiment. Proc Natl Acad Sci 118. https://doi.org/10.1073/PNAS.2012649118
36. Goddard WA III, Song J (2023) Grand canonical quantum mechanics with applications to mechanisms and rates for electrocatalysis. Top Catal 66:1171–1177
37. Melander MM, Wu T, Weckman T, Honkala K (2024) Constant inner potential DFT for modelling electrochemical systems under constant potential and bias. NPJ Comput Mater 10(5)
38. O'M Bockris J, Kahn SUM (1993) Surface electrochemistry: a molecular level approach. Plenum, New York
39. Santos E, Schmickler W (2023) The potential of zero charge of a metal electrode and the surface potential of water from simulations. Curr Opin Electrochem 38(101208):1–8
40. Tian F, Anderson AB (2011) Effective reversible potential, energy loss, and overpotential on platinum fuel cell cathodes. J Phys Chem C 115:4076–4088
41. Dauenhauer PJ (2023) Up up down down left right left right B A start for the catalytic hacker of programmable materials. Matter 6:4145–4157
42. Huang J (2023) Density-potential functional theory of electrochemical double layers: calibration on the Ag(111)-$KPF_6$ system and parametric analysis. J Chem Theory Comput 19(3):1003–1013
43. Zhang M, Chen Y, Eikerling M, Huang J (2025) Structured solvent on a split electron tail: a semiclassical theory of electrified metal-solution interfaces. Phys Rev Appl 23(27):024009
44. Abidi N, Steinmann SN (2023) An electrostatically embedded QM/MM scheme for electrified interfaces. CS Appl Mater Interfaces 15(20):25009–25017
45. Jinnouchi R, Minami S Machine learning force fields in electrochemistry: from fundamentals to applications. ACS Nano, to be published. https://doi.org/10.1021/acsnano.5c05553
46. Anderson AB (2025) How computational theorists deal sith electrode potential. J Electrochem Soc 172:036501

# Chapter 10
# Advanced First Principles Studies Involving Hydrogen Adsorption

**Abstract** This chapter analyzes recent literature studies, demonstrating the kinds of detailed mechanistic questions that can be addressed with modern theories that invoke surface charging and electrolyte response to attain desired electrode potentials. Theoretical calculations and experimental measurements are found to agree that underpotential deposited (upd) hydrogen atoms begin to form as the potential of a Pt electrode is decreased through the double layer range to 0.35 V (RHE). As the potential is decreased from the double layer range, when 0.05 V is reached, coverage of H(ads) reaches 2/3 ML and $H_2$ evolution commences. The question is addressed whether H(ads) can participate in reduction reactions. Theory shows the reaction energy for an R-H bond dissociating on the Pt surface is zero in the upd H range if sites for R(ads) and H(ads) are available.

## 10.1 General Introduction

Water reduction of to form H(ads) which combine to generate $H_2(g)$ is becoming increasingly important for the world's energy infrastructure because $H_2(g)$ can be a fuel for internal combustion engines or for fuel cells generating electricity. Experimental and theoretical practitioners are continuing to explore catalysts for forming $H_2(g)$ by electroreduction. This section discusses results from different levels of theoretical modeling, including CHE and from more advanced interface models including surface charging with electrolyte and interface polarization that are explained in detail later.

 
A. Anderson, *Past and Current Models in Electrochemistry*,
https://doi.org/10.1007/978-3-031-97791-6_10

## 10.2 Participation of Underpotential Deposited (upd) H(ads) on Pt(111) During Electrode Surface Reactions

The previous chapter treated $O_2$ reduction chemistry as a series of four electron + proton transfers leading to water product. In acid, the protons are provided by hydronium ions and in base water molecules are the proton source. Reversing these steps oxidizes water to $O_2$. Can an alternative mechanism participate, where, for example, hydronium ions are discharged putting H on the surface and the H(ads) combine with the intermediates undergoing reduction? This question was addressed as part of a theoretical study R-H of oxidation [1]. The interface code was employed, including surface charging and double layer response. Before delving into the theoretical results a review of experimental facts follows.

The electrode surface condition and its dependence on the applied potential play a critical role in understanding why reduction can occur by proton discharge directly on the adsorbed molecule. From cyclic voltammograms for Pt(111) electrodes it is known that there is an underpotential deposited potential (upd) range where H(ads) is present at varying coverage depending on the potential. As the potential is increased, the low coverage potential limit in pH = 1 electrolyte is reached at about 0.35 V(SHE) [2–4]. The work Orts et al. and theory [5, 6] showed that the shapes of the voltammogram profiles were dictated by lateral interactions of the H(ads) and the maximum coverage of 2/3 ML occurred at 0.05 V (RHE), which marked the onset of evolution of $H_2$. Theory predicted that at 0.1 V the coverage was ½ ML. Consistent with this, during ethanol oxidation experiments ethanol adsorption was completely blocked at 0.06 V while at potentials above 0.1 V ethanol oxidation was strong, indicating adsorption was no longer blocked [7]. These findings mean that electron transfer reactions involving adsorbates are possible when hydrogen is present on the Pt surface from about 0.1 V to 0.35 V. An interesting question is: When can H(ads) compete with $H^+$(aq) as a source on H atoms for reduction reactions in aqueous electrolytes?

## 10.3 When Can an Adsorbate Be Reduced by $H^+$(aq) + $e^-$, by H(ads), or by Both?

Two reactions with a shared intermediate are expected to be complicated to analyze. Different reactions are expected to have different reversible potentials so that if the potential is scanned, one reaction becomes replaced by the other. If they have the same reversible potential, they will be in equilibrium with one another. This section explores a special case involving hydrogen. On Pt(111) electrodes $H^+$(aq) can be reduced to H(ads) over a ~0.35 V underpotential deposition range, the coverage becoming saturated approaching 0.0 V and zero at 0.35 V. This means there may be active sites for reaction of other species over part of the upd range. Might the dissociation mechanism for an RH bond on an electrode surface be different in the upd H range from the mechanism higher potentials?

The question is addressed by examining the reduction of R(ads) to form RH at reversible potential $U_1$, which may be considered an independent variable and has a different value for each species R. Suppose that at potential $U_1$ between 0.1 V and 0.35 V there are sites where R adsorbs with adsorption energy independent H coverage. Consider the reduction reaction

$$\mathrm{R(ads)} + \mathrm{H}^+\mathrm{(aq)} + \mathrm{e}^-(U_1) \rightleftharpoons \mathrm{RH(ads)} \tag{10.1}$$

Then writing

$$\Delta G_1 = G(\mathrm{RH_{ads}}) - G(\mathrm{R_{ads}}) - G\left(\mathrm{H}^+_{\mathrm{aq}}\right) + 4.43\mathrm{eV} + U_1 F = 0 \tag{10.2}$$

The following equilibrium holds simultaneously:

$$\mathrm{H(ads)} \rightleftharpoons \mathrm{H}^+\mathrm{(aq)} + \mathrm{e}^- \tag{10.3}$$

$$\Delta G_2 = G\left(\mathrm{H}^+_{\mathrm{aq}}\right) - 4.43\mathrm{eV} - U_1 F - G(\mathrm{H_{ads}}) = 0 \tag{10.4}$$

Adding Eqs. (10.1) and (10.3) gives, at electrode potential $U_1$,

$$\mathrm{R\ (ads)} + \mathrm{H\ (ads)} \rightleftharpoons \mathrm{RH\ (ads)} \tag{10.5}$$

Adding Eqs. (10.2) and (10.4) gives the Gibbs energy for forming the equilibrium in Eq. (10.5):

$$\Delta G_3 = \Delta G_1 + \Delta G_2 = 0 \tag{10.6}$$

This means that when $U_1$ is in the UPD H potential range, 0.1 V to 0.35 V, the energy for dissociating the RH(ads) on the surface is zero. Additionally, H⁺(aq) can be reduced to H(ads) and H⁺(aq) can reduce R(ads) to RH(ads). This corresponds to a three-way equilibrium (Fig. 10.1).

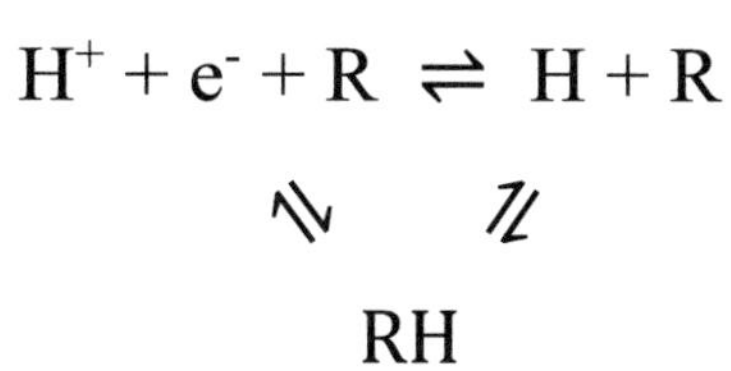


**Fig. 10.1** When the reversible potential $U_1$ for reducing R, Eq. (10.1), is in the upd H range these equilibria hold for adsorbed H, R, and RH when the electrode potential is set to $U_1$

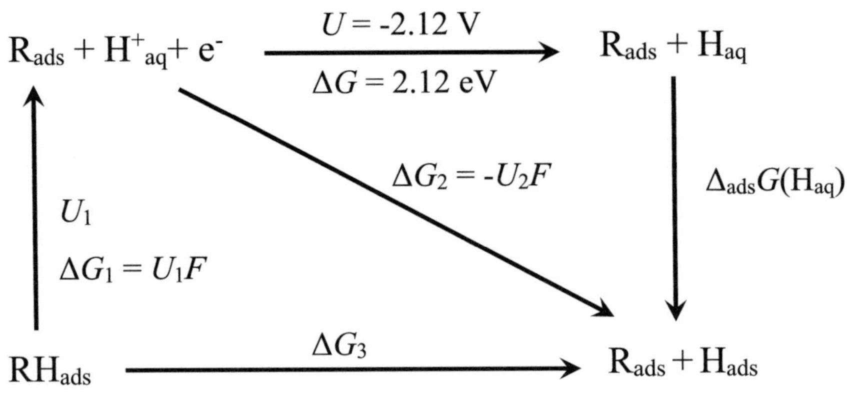


**Fig. 10.2** Pathways for dissociating $RH_{ads}$ on an electrode surface based on Fig. 10.1 in A. B. Anderson and M. Zhao, "Reaction Energy for an Electrode Surface Atom Inserting into an R-H Bond and Its Dependence on Electrode Potential: Application to Pt(111)," J. Electrochem. Soc. 162, H 583-H 589, 2015. This diagram shows three pathways to form R(ads) + H(ads) from RH(ads) when the Pt(111) electrode potential is in the UPD H range. One is RH(ads) dissociation; another is RH(ads) oxidation to R(ads) + $H^+$(aq) + $e^-$ at $U_1$ followed by $H^+$ reduction to H(ads) at $U_2$; and the last one is reducing $H^+$(aq) to H(aq) at −2.12 V and then adsorbing it

A Haber cycle can be constructed to incorporate the potential dependence of the H adsorption energy, $\Delta_{ads}G(H_{aq})$. To do this requires having the value of the standard reduction potential for forming H(aq) by

$$H^{+}(aq) + e^{-} \rightleftharpoons H(aq) \tag{10.7}$$

This was calculated to be −2.12 V in Ref. [1]. Using this equation, $\Delta_{ads}G(H_{aq})$, and Eq. (23) from Chap. 8, Fig. 10.2 is generated.

When the "ads" subscripts are dropped in Fig. 10.2, the process RH(aq) going to R(aq) + H(aq) may be considered. In this case the RH(aq) bond dissociation energy is the sum of two terms, $\Delta G_1$ and 2.12 eV. This is a general equation for determining any RH(aq) bond strength when the oxidation reversible potential is known. Its value in non-aqueous as well as aqueous environments will be subject to relatively small corrections for solvation energies since the reactant and products of the dissociation are uncharged.

As the applied potential $U$ is increased through the UPD potential range, the H adsorption energy decreases, meaning the adsorption bond becomes stronger because of the reduction in through bond repulsion as the H(ads) coverage decreases. This means the relation between free energy of adsorption and coverage follows a Frumkin isotherm and not a Langmuir isotherm where adsorption is randomized over the surface because adsorbates do not interact [8]. Assuming there is linear dependence of adsorption energy on potential, one can write for potential range 0.0 V to 0.35 V:

$$\Delta_{ads}G(H_{aq}) = -2.12\text{eV} - UF \tag{10.8}$$

At 0.35 V, $\Delta_{ads}G(H_{aq})$ is −2.47 eV and this value is assumed unchanged at potentials greater than 0.35 V going through the double layer range up to ~0.6 V where OH(ads) begins to form by water oxidation [9]. The slow potential dependence of the PtH bond strength at low coverage on clean Pt(111) was shown elsewhere [10]. Over the UPD hydrogen potential range, the equalities in Eq. (10.6) hold but when the Pt(111) electrode potential is increased positive of the upd H range into the double layer potential region, $\Delta G_3$ is no longer zero. At these potentials it increases linearly with the potential because H(ads) is more stable when oxidized and removed as $H^+(aq)$; and it costs energy to keep it in the reduced state. The increase continues through the double layer potential range to about 0.6 V. If RH molecules with oxidation reversible potentials $U_1$ between 0.35 V and 0.60 V are able to displace adsorbed water molecules at the surface to become oxidized, they have must dissociation energies greater than zero:

$$\Delta G_3 = (-0.35 + U_1/F)\text{eV} \quad (U_1 > 0.35\text{V}) \tag{10.9}$$

Water is oxidized to OH(ads) above 0.6 V, blocking the surface which changes its adsorption properties. In the other direction, H(ads) saturation is reached and when the potential is less than 0.0 V and dissociation of any RH(ads) that forms becomes exergonic with energies

$$\Delta G_3 = U_1 F \tag{10.10}$$

However, H(ads) and $H_2$ bubbles should sufficiently block the surface to prevent RH from adsorbing.

At potentials greater than 0.35 V RH dissociation is limited by its activation energy and by the increase in Gibbs energy for forming R(ads) + H(ads). When RH is oxidized on the surface at potential $U_1 > 0.35$ V, the H(ads) will be oxidized off. However, deprotonation into solution will be an increasingly competitive pathway for RH oxidation as the potential increases. Figure 10.3 sums up this discussion winch is expected to also apply for alcohols.

Dissociation of an OH bond in $H_2O$(ads) will also be endergonic at potentials greater than 0.35 V. This implies that OH(ads) formation, which commences at about 0.6 V, should not be by the dissociative mechanism. This leaves deprotonation into solution as the probable mechanism. As pointed out in Chap. 8, local reaction center (LRC) calculations suggest low activation barriers to this. Assuming these conclusions are correct, H(ads) will not exist on Pt(111) at potentials above 0.6 V and modeling without taking potential into account, which is sometimes done, misses this [11, 12].

The investigations of Orts et al. and others [13] found the (110) and (100) surfaces of platinum have H upd and double layer ranges different from that of Pt(111). Other metals have still different properties. At this time, it is not known how much the picture presented here can be generalized to other surfaces.

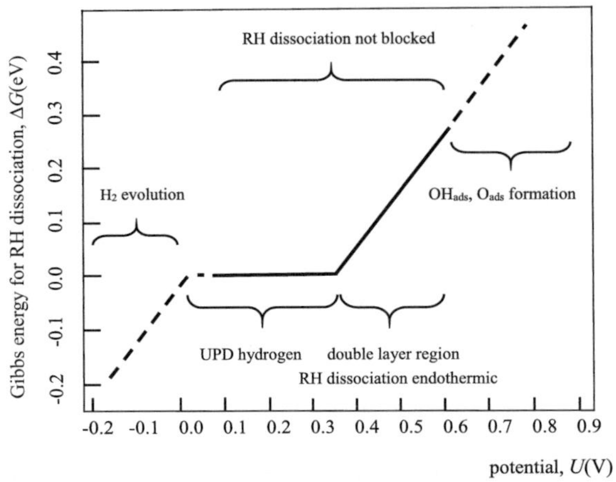


**Fig. 10.3** Graph of the reaction energy for RH dissociation to R(ads) + H(ads) at a Pt(111) electrode surface in standard electrolyte and also showing the condition of surface a function of applied potential. The reaction energy is zero over the UPD hydrogen potential range from 0 V to 0.35 V and it increases lineally with potential in the double layer range from 0.35 V to 0.6 V, above which the water oxidation reaction blocks RH dissociation. It decreases in the $H_2$ evolution range below 0.0 V but is blocked by hydrogen evolution. Based on Fig. 5 in A. B. Anderson and M. Zhao, "Reaction Energy for an Electrode Surface Atom Inserting into an R-H Bond and Its Dependence on Electrode Potential: Application to Pt(111)," J. Electrochem. Soc. 162, H 583-H 589, 2015

## 10.4 Participation of Overpotential Deposited (opd) H(ads) on Cu(111) During Electrode Surface Reactions

Compared to platinum, copper is notable for its absence of upd hydrogen. Sharifi-Asi and Macdonald found $H_2$ evolution occurs at negative potentials following overpotential deposition (opd) of H [14]. They measured steady state polarization curves using rotating polycrystalline Cu disk electrodes at pH 5.72, 8,00, and 9.20. Tiny current densities began at about −0.2 V (RHE) and rose to the 0.1 mA/cm$^2$ level at −0.5 V (RHE) and rose rapidly at increasingly negative potentials. An Arrhenius activation energy of about 0.33 eV was estimated using the dependence of measured exchange current densities on temperature. The log($i_0$) versus $1/T$ had broad data scatter.

DFT calculations including surface charging and electrolyte polarization show the reaction

$$H_2(g) \rightarrow 2H(ads) \tag{10.11}$$

is endothermic by about 0.25 eV on Cu(111) under standard conditions at the pzc [15]. The above measurements showed that $H_2$ evolution is very slow at 0 V (RHE) and starts to increase at around −0.2 V (RHE) on the μA/cm$^2$ scale, that is, at about −0.2 V overpotential. With the goal of understanding the surface conditions, H(ads) coverage from hydronium ion discharge on the surface was calculated as a function of potential. Theory predicted almost zero H(ads) coverage at 0 V (RHE) and about 1/3 ML coverage at −0.2 V (RHE) and about 0.9 ML coverage at −0.5 V (RHE). Coverage-dependent reversible potentials were calculated using Gibbs energies of the left- and right-hand sides of the reaction

$$\mathrm{H(ads,\,\Theta) + H(aq) + e^{-}}(U) \rightleftharpoons \mathrm{H(ads,\,\Theta + \Delta)} \tag{10.12}$$

for several values of coverage, Θ. Similar potential-dependent coverages will hold for less acidic and basic studies as was found in the work of Sharifi-Asi and Macdonald. The H adsorption energy at the pzc was predicted to decrease at a rate of 0.27 times the rate for increase in coverage. Based on the low calculated barriers to surface diffusion, mobility of H(ads) at the pzc was expected to be high on Cu(111), allowing combination of two H(ads) by the Tafel mechanism to form $H_2$(g). Assuming the Volmer-Tafel mechanism, the observed increase in the $H_2$ evolution rate as the applied potential was made increasingly negative could be explained by the high mobility and weaker bonding if H to the surface at increasing coverage.

Sharifi-Asi and Macdonald extended their kinetic model to include data from electrochemical impedance spectroscopy (EIS) to interpret parameters for (i) $OH^-$ adsorption, $H_2O$ discharge forming H(ads) in a Volmer step, (ii) $H_2O$ discharge over H(ads) forming $H_2$(g) in a Heyrovsky step, and (iii) two H(ads) reacting to form $H_2$(g) in a Tafel step. They suggested the Volmer step was probably rate limiting and both Volmer-Heyrovsky and Volmer-Tafel mechanisms participated in $H_2$(g) evolution.

Another experimental study of $H_2$ evolution centered on the interest of using copper electrodes for electrocatalyzing the reduction of $CO_2$ by hydrogen [16]. Reducing of $CO_2$ is an ongoing research topic because of contemporary interests in remediation atmospheric $CO_2$ and using $CO_2$ as a feedstock for producing chemicals. More information about hydrogen evolution comes from rotating polycrystalline disk Cu electrode measurements at different pH values [16]. In Ref. [16] the authors found rapid increase in current density on the mA/cm$^2$ scale beginning around −0.5 V (RHE) during negative scanning of the potential. This became limited by the $H^+$(aq) diffusion rate at about −0.6 V (RHE) for 400 rpm and at about −0.7 V (RHE) for 2500 rpm. At around −1.2 V (RHE) the current density began increasing rapidly from this $H^+$(aq) diffusion limited plateau and this increase was attributed to $H_2$ evolution from water reduction. The hydrogen evolution (HER) pathway could not be fully elucidated, that is, whether it involved discharge of $H_2O$ forming H(ads) by the Volmer reaction, followed by combination with a second H(ads) via the Tafel reaction to form $H_2$(g) or by the Heyrovsky reaction, discharging $H_2O$ over H(ads) to form $H_2$, or a combination of both HER mechanisms. FTIR spectral measurements indicated that when present, $CO_2$ can be reduced forming CO(ads) which inhibited hydrogen evolution.

A 2021 study compared $H_2$ evolution and $CO_2$ reduction reactions on metallic Cu and $Cu^0$ present on partially reduced surfaces of $Cu_2O$ [17]. The computational method used included surface charging and model electrolyte response to achieve desired potentials. Potential-dependent interface polarization was included, putting the calculational method in the advanced category. The experiments showed $CO_2$ reduction in saturated pH 7 $K_2CO_3$ at $-1.0$ V (RHE) over $Cu_2O$ electrode surfaces initially produced $H_2$(g) with ethylene and small amounts of methane but after several hours $H_2$(g) became the exclusive product. DFT modeling supported an argument that the surface Cu associated with sub-surface O were sites the OCCO intermediate bonded more strongly to than the Cu(111) surface. Over the active surfaces the reaction energy was nearly energy neutral compared to Cu(111) and activation energies were lowered. The HER mechanism was explored computationally and was predicted to begin with the Volmer step, $H_2O$ discharge forming H(ads) + $OH^-$(aq) followed by the Heyrovsky step, with a second $H_2O$ discharge over H(ads), forming $H_2$(g) + $OH^-$(aq). The calculations were performed on Cu(111) at $-1.0$ V (RHE) and the rate determining step (RDS) was Heyrovsky, with activation energy 0.79 eV, which was 0.16 eV lower than for CC coupling. For $Cu^0$ the RDS was the Volmer reaction, with activation energy 0.70 eV, which was 0.02 eV lower than for CC coupling. Missing are consideration of the alternative H–H coupling mechanism of two H(ads) by the Tafel reaction. Also missing is the weakening effects of high H(ads) coverage at this potential (Frumkin effect) as found in the theoretical predictions of Zhao and Anderson [15]. The activation energy for the Tafel mechanism will decrease as the potential and coverage become more negative but the role of this as an alternative mechanism was not explored in the publication.

Recently electrochemical mass spectrometry was combined with chronoamperometry to quantify hydrogen coverage on Cu(111) in sulfuric acid electrolytes [18]. The authors determined that at potentials increasingly negative of 0 V, H coverage increased to a maximum of 2/3 ML at –0.4 V and the analysis indicated H(ads) recombination was the mechanism of $H_2$(g) formation. These are also the conclusions of the theoretical study of Zhao and Anderson in 2015, [15] which was outlined above.

## 10.5 Hydrogen Adsorption by the CHE Method

Let us consider CHE predictions of potential-dependent reaction energies from a broader perspective. One viewpoint begins with a Taylor series in surface energy in terms of excess surface charge in a fully grand-canonical approach, meaning the system is open to changes in energy and the number of particles. In the expansion the zeroth-order component of the energy is the CHE energy [19]. In CHE the potential is introduced only through the electron energy and the electron comes from a non-interacting donor energy level that is part of the system, as in the Gurney picture. This feature is evident in the way the following formula is applied

$$\mu(H^+) + \mu(e^-) = 1/2\mu(H_2) - eU \tag{10.17}$$

The interpretation of Eq. (10.17) is that on the standard hydrogen electrode (SHE) scale, at 0 V, $\mu(H^+) + \mu(e^-) = 1/2\mu(H_2)$ and an energy $eU$ for an electron may be defined such that it is 0 when the electrode is at 0 V (SHE) and is assigned values $eU$ when the electrode potential is greater than or less than 0 V (SHE). However, in the CHE model the electrode surface charge is zero and the surface workfunction, Fermi level, and potential are dependent only on whatever is adsorbed and not on the potential assigned to the electron. Having the electron originate from a source in the system with controlled potential was also used in the LRC method introduced in Chap. 8 for predicting transition states and activation energies.

The CHE method was extended by Chan and Nørskov to include transition states during the course of the $H_2$ generation reactions [20]. In the model the net charge on translational cells used in the interface band calculations was zero. A hydrogen atom was bonded to a water molecule associated with an ice-like water bilayer above the surface metal slab translational cells. The H atom spontaneously lost an electron to the metal, forming hydronium ion. Using a double layer capacitor model, the authors extrapolated to the low coverage limit and calculated reaction energies and activation energies for the Volmer, Heyrovsky, and Tafel reactions. The total energy $E$ consisted in two terms:

$$E = E_{\text{chem}} + E_{\text{el}} \tag{10.18}$$

The $E_{\text{chem}}$ contribution is the DFT "chemical" energy calculated variationally for initial, transition, and final states; $E_{\text{el}}$ is an electrostatic energy for the capacitor interface. Due to charge delocalization, unit charges were not relevant and the Bader charge separation model was applied to the orbitals to assign charges for the capacitor model. Two-parameter charge extrapolation formulas were used to assign adsorption energies and transition state energies for H. Results were presented for $H_2$ and $H_2O$ formation on Pt(111).

A following paper gave additional results of extended CHE calculations modeling effects of changing potential and coverage of H(ads) [21] Activation energies, Tafel plots, and polarization curves were calculated but how much they are in error due to the model remains to be determined. The conclusion that electrochemical $H_2$ generation follows the Volmer-Heyrovsky pathway over the metal surfaces considered may be correct but more physically realistic models should be applied such as in the work of G. Liu et al. [17]. In the case of Cu(111) the Volmer-Heyrovsky pathway was predicted favored because the calculated activation energy for the Heyrovsky step at 0.9 ML H(ads) coverage was lower than for the Tafel step, which is in opposition to the conclusion of G. Liu et al. For Cu(111) the recent work of Raciti and Moffat that was mentioned above [18] showed recombination of opd H(ads) surface layers was a viable mechanism for $H_2$ formation when the electrode potential was shifted in the positive direction. During steady state $H_2$ evolution at fixed potential the jury is still out concerning the involvement of the Heyrovsky step.

## References


1. Anderson AB, Zhao M (2015) Reaction energy for an electrode surface atom inserting into an R-H bond and its dependence on electrode potential: application to Pt(111). J Electrochem Soc 162:H 583–H 589
2. Gomez R, Orts JM, Alverez-Ruiz B, Feliu JM (2004) Effect of Temperature on hydrogen adsorption on Pt(111), Pt(110) and Pt(100) electrodes in 0.1 M $HClO_4$. J Phys Chem B 108:228–238
3. Strmcnik D, Tripkovic D, van der Vliet D, Stamenkovic V, Markovic NM (2008) Adsorption of hydrogen of Pt(111) and Pt(100) surfaces and its role in the HOR. Electrochem Comm 10:1602–1605
4. Jerkiewicz G (2010) Electrocatalytic hydrogen adsorption and absorption. Part 1: under-potential deposition of hydrogen. Electrocatal 1:179–199
5. Asiri HA, Anderson AB (2013) Using Gibbs energies to calculate the Pt(111) Hupd cyclic Voltammogram. J Phys Chem C 117:17509–17513
6. Karlberg GS, Jaramillo TF, Skulason E, Rossmeisl J, Bligaard T, Nørskov JK (2007) Cyclic Voltammograms for H on Pt(111) and Pt(100) from first principles. Phys Rev Lett 99:126101–1–4
7. Wang Q, Sun GQ, Jiang LH, Xin Q, Sun SG, Jiang YX, Chen SP, Jusys Z, Behm RJ (2007) Adsorption and oxidation of ethanol on colloid-based Pt/C, PtRu/C and $Pt_3Sn$/C catalysts. In Situ FTIT spectroscopy and on-line DEMS studies. Phys Chem Chem Phys 9:2686–2696
8. Gileadi E (1993) Electrode kinetics for chemists, chemical engineers, and materials scientists. Wiley-VCH, Ney York
9. Rizo R, Sitta E, Herrero E, Climent V, Feliu JM (2015) Towards the Understanding of the Interfacial pH Scale at Pt(111) Electrodes. Electrochim Acta 162:138–145
10. Jinnouchi R, Anderson AB (2007) Density functional theory/continuum model approach to surface electrochemistry. COLL 409, 234th ACS Meeting, Boston MA
11. Sha Y, Yu TH, Metinov BV, Shrinivanian P, Goddard WA III (2012) Mechanism for oxygen reduction reaction on a $Pt_3Ni$ alloy fuel cell cathode. J Phys Chem C 116:21334–21342
12. Duan Z, Wang G (2013) Comparison of reaction energetics for oxygen reduction reactions on Pt(100), Pt(111), Pt/Ni(100), and Pt/Ni(111) surfaces: a first-principles study. J Phys Chem C 117:6284–6292
13. Garcia-Araez N (2011) N. Enthaplic and entropic effects on hydrogen and OH adsorption on Pt(111), Pt(100), and Pt(110) electrodes as evaluated by gibbs thermodynamics. J Phys Chem C 115:501–510
14. Sharifi-Asi S, Macdonald DD (2013) Investigation of the kinetics and mechanism of the hydrogen evolution reaction on copper. J Electrochem Soc 160:H382–H391
15. Zhao M, Anderson AB (2017) Theory of hydrogen deposition and evolution on Cu(111) Electrodes. J Electrochem Soc 164:H691–H695
16. Ooka H, Figueiredo MC, Koper MTM (2017) Competition between hydrogen evolution and carbon dioxide reduction on copper electrodes in mildly acidic media. Langmuir 33:9307–9313
17. Liu G, Lee M, Kwon S, Zeng G, Eichhorn J, Buckley AK, Toste FD, Goddard III WA, Toma FM (2021) $CO_2$ reduction on pure Cu produces only $H_2$ after subsurface O is depleted: theory and experiment. https://doi.org/10.1073/pnas.2012649118
18. Raciti D, Moffat TP (2022) Quantification of hydride coverage on Cu(111) by electrochemical mass spectrometry. J Phys Chem C 126:18734–18743
19. Hormann NC, Marzari N, Reuter K (2020) Electrosorption at metal surfaces from first principles. npj Comput Mater 136:1–10
20. Chan K, Nørskov JK (2015) Electrochemical barriers made simple. J Phys Chem Lett 6:2663–2668
21. Tang MT, Liu X, Ji Y, Nørskov JK, Chen K (2020) Modeling hydrogen evolution reaction kinetics through explicit water-metal interfaces. J Phys Chem C 124:28083–28092

# Chapter 11
# First Principles Modeling at the pzc Including Polarizable Electrolyte: Oxygen Reactions on the Pt Cathode and Electroless Copper Deposition

**Abstract** Principles introduced in recent literature are discussed. Based on pzc calculations of exergonic dissociation energies for OOH(ads) forming O(ads) + OH(ads) using the interface code, it is shown that the energy loss leads to an effective reversible potential ($U_{\text{eff}}$) and consequently the observed overpotential during oxygen reduction. An ideal catalyst has no energy loss for the dissociation or any other steps during the four-electron reduction. At 1.23 V the adsorption energies on an ideal catalyst are, in the pzc model, 0.0, 1.35, 2.41, 1.37, and 0.0 eV for $O_2$, OOH(aq), O(aq), OH(aq), and $H_2O$(ads) respectively. Using the simplified LGER model for platinum-based catalysts at the pzc, including Pt skins on Pt alloys, it is concluded that bond strengths of OH and O are high and expected to control $U_{\text{eff}}$. These two bond strengths are referred to as descriptors in the literature. Turning to an autocatalytic reaction, mechanisms and understanding for electroless deposition of copper are proposed. Predictions are compared with experiment and the application of mixed potential theory. The results indicate the generally assumed mixed potential theory does not explain the details of electroless deposition of copper.

## 11.1 General Introduction

As shown in Chap. 8, the CHE and LGER methods have enjoyed and continue to enjoy continuing application using DFT codes without inclusion of the electrolyte in the calculations. This chapter covers results from applications of the interface code, including a polarizable electrolyte with its distribution at the interface determined using a modified Poisson-Boltzmann (MPB) theory, but they are at the pzc. This allows the definition of effective reversible potential ($U_{\text{eff}}$) using high quality calculations. Oxygen reduction over Pt(111) electrodes is treated and $U_{\text{eff}}$ is the consequence of energy loss when the first reduction intermediate OOH(ads) dissociates on the surface. Interface calculations show the pzc for a given adsorbed reaction intermediate is dependent on its structure and throughout the course of a multistep reaction the pzc will wander. While the first step in $O_2$ reduction is usually thought to



A. Anderson, *Past and Current Models in Electrochemistry*,
https://doi.org/10.1007/978-3-031-97791-6_11

be OOH(ads) formation followed by its dissociation, pzc calculations for Pt monolayer skins on 12 transition metals suggest uncertainty and dissociative adsorption of $O_2$ with subsequent electron transfers may be possible over some surfaces. Studies with variable electrode potential should sort out the two mechanisms in the future. The interface and similar codes can also be used to address what are essentially organometallic redox mechanisms. This is demonstrated by analysis of electroless deposition of copper also based on pzc calculations. Chapter 12 will introduce results for controlled changes of the electrode potential using the interface code applied to $H_2O$ oxidation and H deposition.

## 11.2 Defining the Thermodynamically Ideal Oxygen Reduction Electrocatalyst from pzc Calculations Including the Polarizable Double Layer

Many standard reversible potentials for bulk solution reactions are known experimentally [1–3]. In this section reactions oxidizing $O_2$(g) to $H_2O$(l), which is the cathode reaction in fuel cells, are used to demonstrate the application of the LGER model and deduce general conclusions. Platinum is the standard catalyst used for the cathode reaction and there are continuing efforts to enhance cathode activity by alloying platinum or replacing it with less expensive materials.

Consider the reaction mechanism in solution to have four steps as follows:

$$\mathrm{O_2(g) + H^+(aq) + e^- \rightleftharpoons OOH(aq)},\ U^0 = -0.125\mathrm{V},\ \Delta G = 0.125\mathrm{eV} \tag{11.1}$$

$$\mathrm{OOH(aq) \rightarrow O(aq) + OH(aq)},\ \Delta G = 2.52\mathrm{eV} \tag{11.2}$$

$$\mathrm{O(aq) + H^+(aq) + e^- \rightleftharpoons OH(aq)},\ U^0 = 2.12\mathrm{V},\ \Delta G = -2.12\mathrm{eV} \tag{11.3}$$

$$\mathrm{2OH(aq) + 2H^+(aq) + e^- \rightleftharpoons H_2O(aq)},\ U^0 = 2.72\mathrm{V},\ \Delta G = -5.44\mathrm{eV} \tag{11.4}$$

The reversible potentials $U^0$ are sourced from the references and the sum of $\Delta G$ for the reactions is −4.92 eV, corresponding to 1.23 V for the four-electron process:

$$\mathrm{O_2(g) + 4H^+(aq) + 4e^- \rightleftharpoons 2H_2O(aq)} \tag{11.5}$$

One could attempt to perform electron transfer steps of Eqs. (11.1), (11.3), and (11.4) separately over an electrode at its own reversible potential or at a potential negative of it. But together on an electrode each reaction will have to take place at a single electrode potential. Removing the reduction intermediates as they form at their reversible potentials and placing them in cells running at different potentials is not possible. Also, finding a way to dissociate OOH(aq) is a complication.

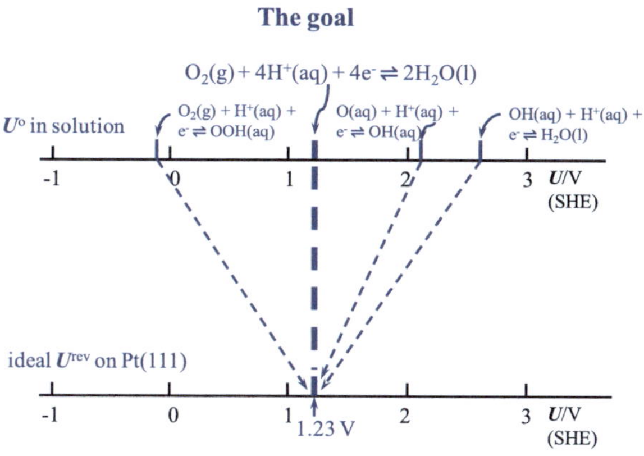


**Fig. 11.1** Diagram with two parallel potential scales spanning −1 to 3 V. The upper one is marked at the experimental standard reversible potentials for $O_2$ reduction to OOH, O reduction to OH, and OH reduction to $H_2O$. The lower one is marked at 1.23 V, the ideal reversible potential for oxygen reduction to water on Pt(111), and three dashed lines connect this with the three marks on the upper scale. Reproduced from Ref. [4] with permission from the Royal Society of Chemistry

An electrocatalyst having the reversible potential for each electron transfer step in Eq. (11.5) shifted to 1.23 V by the reactant and product adsorption energies would be thermodynamically ideal. Additionally, for no Gibbs energy to be used or lost, the OOH(ads) dissociation energy forming O(ads) + OH(ads) on the surface must be 0.00 eV. The thermodynamically perfect catalyst will accomplish all these feats as illustrated in Fig. 11.1 [4].

The catalyst must shift the potential for $O_2$(g) reduction to OOH(aq) 1.36 V positive and for O reduction to OH the shift is 0.89 V negative and for the potential for its reduction to $H_2O$ the shift is 1.29 V negative. According to LGER, all electron transfer steps will be at 1.23 V when the OOH(ads) dissociation Gibbs energy is 0.0 eV and the respective adsorption Gibbs energies for $O_2$, OOH(aq), O(aq), OH(aq), and $H_2O$(ads) are 0.0, 1.35, 2.41, 1.37, and 0.0 eV. These are the adsorption properties of the thermodynamically ideal oxygen reduction electrocatalyst as based on pzc calculations including a polarizable double layer. It can be shown that these adsorption energies apply for other $O_2$ reduction mechanisms as well, including the initial dissociative $O_2$ adsorption pathway and pathways where OOH(ads) is reduced to O(ads) + $H_2O$(l) or $H_2O_2$, which may either dissociate on the surface or be reduced to OH(ads) + $H_2O$(l).

An ideal electrocatalyst as defined above is ideal in terms of thermodynamics. Even if the reaction energies were ideal, the reaction rates will still depend on the activation energies for O–O bond dissociation and the electron + proton transfer steps. Also, the $O_2$(g) and $H_2O$(l) adsorption energies are expected to be small but not exactly zero. Nevertheless, over a "perfect" electrocatalyst the reduction potentials for the three electron transfer steps will be 1.23 V, the activation energies for the steps will all be zero, and $O_2$ adsorption and $H_2O$ desorption will not release or absorb Gibbs free energy.

## 11.3 Energy Loss and Effective Reversible Potentials ($U_{eff}$)

If during the four-electron reduction to water OOH(ads) undergoes exergonic dissociation on the catalyst surface, that amount of Gibbs free energy is lost, dissipated as heat and entropy, and not available for performing electrical work. The concept of an effective reversible potential was introduced as the consequence of the energy loss [5]. In this reference surface charging with modified Poisson-Boltzmann (MPB) polarizable double layer theory was employed. The OOH(ads) dissociation Gibbs energies calculated at the pzc for Pt(111) surfaces covered with different amounts of OH(ads) and $H_2O$(ads) were all close to –1.2 eV. Then energy available for the four-electron reduction decreases from 4(1.23 eV) = 4.92 eV to 3.7 eV and the reversible potential, which, $U_{eff}$, is then 3.7/4 V = 0.93 V. $U_{eff}$ determines the reduction current cut-off in $O_2$ reduction polarization curves over platinum electrodes as depicted in Fig. 11.2, with the current density approaching zero around 0.9 V.

As shown if Fig. 11.2 the current density from 0.5 V to about 0.7 V has a high absolute value and the reaction controlled by the diffusion rate of reactants to the anode surface. From 0.7 to 0.9 V the driving force is less and the current density decreases because of kinetic control, meaning increases in activation energies for some reduction steps slow the overall reaction. At $U_{eff}$, the current density rapidly approaches zero as the potential approaches 1.2 V. Figure 11.3 defines these areas and Fig. 11.4 shows the shifts in reversible potentials calculated for the Pt surface from the standard bulk solution values.

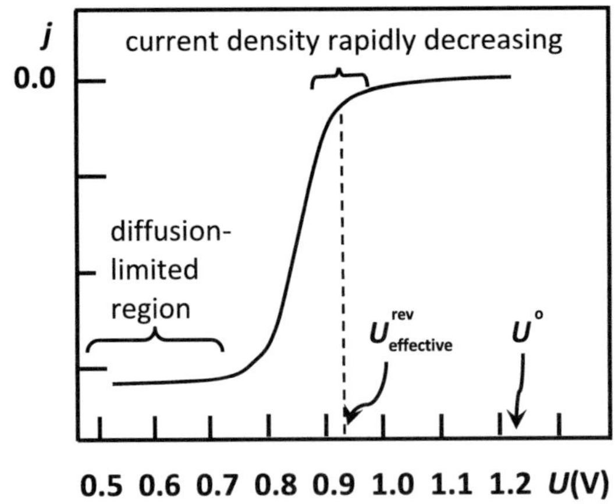


**Fig. 11.2** Schematic s-shaped polarization curve as typically measured from 0.5 to 1.2 V depicting potential-dependent current densities for $O_2$ reduction to water over platinum electrodes. The diffusion limited high current density extends to about 0.7 V followed by a rapid decrease approaching zero at $U_{eff}$, = 0.93 V. Reprinted with permission from F. Tian and A. B. Anderson, "Effective Reversible Potential, Energy Loss, and Overpotential on Platinum Fuel Cell Cathodes," J. Phys. Chem. C 115, 4076–4088, 2011. Copyright 2011 American Chemical Society

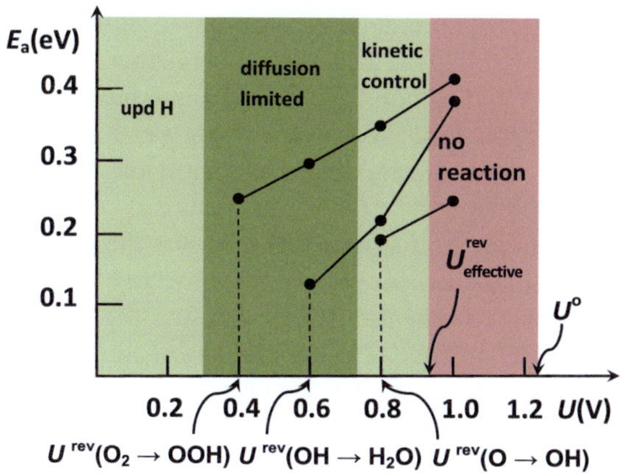


**Fig. 11.3** Potential regions where four-electron reduction of $O_2$ is allowed and under diffusion and kinetic control up to $U_{eff}$ and $U^{rev}$ for the steps shown, as calculated using the interface code. As marked, respective reversible potentials for $O_2$ to OOH, OH to $H_2O$, and O to OH reductions are about 0.4, 0.6, and 0.8 V. upd H is present from 0 V to about 0.3 V; the reaction is diffusion limited from 0.3 V to about 0.75 V; it is under kinetic control from 0.75 V to about 0.93 V. Approximate activation energies are on the vertical scale. Reprinted with permission from F. Tian and A. B. Anderson, "Effective Reversible Potential, Energy Loss, and Overpotential on Platinum Fuel Cell Cathodes," J. Phys. Chem. C 115, 4076–4088, 2011. Copyright 2011 American Chemical Society

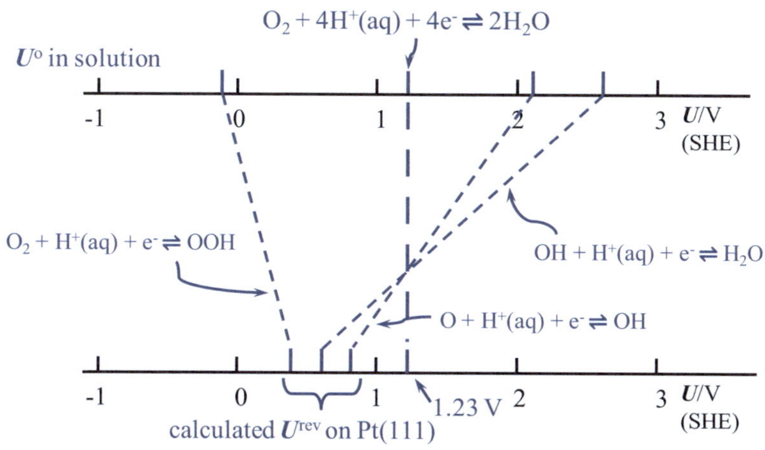


**Fig. 11.4** As in Fig. 11.1 with the calculated reversible potentials for the three steps on Pt(111) all shown to be the values in Fig. 11.3, all less than 1.23 V. [4] Reproduced from Ref. with permission from the royal society of chemistry

The challenge for the ideal $O_2$ reduction catalyst is to shift all reversible potentials in solution to 1.23 V by providing the capability to break the O–O bond with no energy input and no energy loss and to bond the reaction intermediates with the energies listed in the previous section. Finding or engineering a catalyst that will accomplish this goal is a formidable challenge and might be impossible. In the earlier discussion of early applications of semiempirical theory the focus was on increasing the potential of OH(ads) removal from Pt. At that time OH(ads) was thought to "poison" (jargon from chemical catalysis studies) the surface by blocking reaction sites. Alloying Pt with more electropositive transition metals shifted the reversible potential for HO(ads) reduction to $H_2O$ positive. The weakly adsorbed $H_2O$ molecules could be desorbed or displaced, opening up sites for $O_2$ to adsorb and initiate the cathode reaction, and the theory indicated this was because of weakened OH bonding to the alloyed surfaces. However, as Figs. 11.3 and 11.4 show, all reduction steps must be considered as must the energy for O–O(ads) bond breaking.

For maximum power generation the potential of operating hydrogen fuel cells using platinum electrodes is in the 0.6–0.7 V range and a combination of kinetic factors for the steps is responsible. When the potential is pushed toward 0.9 V current density and power decrease asymptotically to zero. A simple assumption is that if catalyst modifications push $U_{\mathrm{eff}}$ closer to 1.23 V the operating potential and power output will increase, but reality is complicated because the surface that accomplishes this must at the same time satisfy the reaction energy and adsorption bond strength constraints in Sect. 11.2.

## 11.4 Possible Inconsistencies Between Figs. 11.3 and 11.4 and the Empirical Fig. 11.2: Is the First Step $O_2$ Dissociation?

It may be asked whether potentials in Figs. 11.3 and 11.4 for Pt (111) are consistent with the polarization curve in Fig. 11.2. The first step, forming OOH(ads) from $O_2$ at about 0.4 V is exergonic by 0.4 eV, OOH dissociation is exergonic by 1.3 eV, O reduction to OH is exergonic by 0.8 Ev, and the two OH reductions to $H_2O$ are exergonic by 1.2 eV. The sum of these energies, –3.7 eV, when divided by 4 V/eV does, in confirmation, give $U_{\mathrm{eff}} = 0.93$ V.

On Pt(111) the calculated potential of OOH(ads) formation is ~0.5 V negative of $U_{\mathrm{eff}}$ and the OH(ads) reduction potential is ~0.3 V negative of it. Are these values inconsistent with reaching the diffusion limited current initiating at about 0.7 V since the activation energy for OOH(ads) formation will, according to Tafel, be increasing as the potential is increased from 0.4? The OH(ads) reduction potential has strong dependence on its coverage as will be discussed in detail later. However, the potential for reducing $O_2$ to OOH(ads) remains around 0.4 V for the various coverages of $H_2O$(ads) and OH(ads) as calculated by Tian and Anderson [5]. The question is answered in Sect. 11.6 which refers to an advanced DFT study that supports high activity for forming OOH(ads) despite the increasing activation energy at potentials greater than 0.4 V.

## 11.5 Consequences of Assuming Different Sites for Different Steps During $O_2$ Reduction

If one optimizes the adsorption energies for one of the reduction reactions so that its reversible potential is 1.23 V, the other steps cannot be expected to also have this optimal reversible potential. Why not? Suppose the catalyst presents four separate sites, possibly adjacent, where on one the OOH(ads) formation potential is 1.23 V, on another OOH(ads) or $O_2$(ads) dissociation energies are zero, on another where O(ads) to OH(ads) reduction potential is 1.23 V, and on another the OH(ads) to $H_2O$(aq) reduction potential is 1.23 V. Keep in mind the ideal electrocatalyst for this reaction has specific adsorption energies for each intermediate. To carry out the four-electron reduction over such a catalyst, the intermediates must move from one site to another and this will either cost energy, either by the need to boost the intermediate uphill to the next site, or by its releasing and wasting energy while sliding down to the next reduction site. Therefore, the ideal catalyst will be a single site with the adsorption energies listed above and with no need for the intermediates to move from the sites where they formed. No catalyst with these energies has been identified and probably none ever will because of the constraints on the adsorption energies of the intermediates. Platinum alloys are the best catalysts found so far for use in fuel cells because of their activity and stability. The following section compares calculated values of $U_{eff}$ for several platinum alloys with Pt skin surfaces.

## 11.6 Calculated $U_{eff}$ for $O_2$ Reduction Over Pt Skins on Transition Metals and O and OH Adsorption Energies as Descriptors. Comparing O–OH and O–O Dissociation Mechanisms Using LGER

Here $U_{eff}$ are calculated for two pathways, OOH(ads) dissociation and $O_2$(ads) dissociation over transition metal surfaces of different compositions. Calculated adsorption Gibbs energies for $O_2$, OOH, O, OH, and $H_2O$ on 12 Pt(111) skins and Au(111) were used to establish trends in $U_{eff}$ for $O_2$ reduction by the two O–O dissociation mechanisms [6]. The energies were obtained at the pzc for use in the LGER model, and the polarizable double layer was included in the DFT calculations. This approach should give useful trends. In this LGER work $U_{eff}$ on Pt(111) is 0.82 V, different from 0.93 V reported in Ref. [5] because the structure models.

Considering the pathway where OOH(ads) is an intermediate and looking at Fig. 11.5, it is seen that the results are ordered so that both the calculated adsorption

Gibbs and the calculated for O adsorption decrease across the series of Pt skins. The OH adsorption Gibbs bond strengths trend the same as those for O but show some deviations. The OOH Gibbs adsorption bond strengths also decrease across the series, but at a lower rate. For the OOH(ads) formation mechanism the consequence is that $U_{eff}$, which depends on the exergonic OOH(ads) dissociation energy generally increases across the series as seen in Table 11.1. The same holds for $U_{eff}$ calculated using the energy loss during exergonic $O_2$(ads) dissociation. The adsorption energies of either OH or O may be employed as descriptors because, according to Fig. 11.5, when one decreases so does the other, leading to increased $U_{eff}$. As the figure shows, the internal energies of adsorption trend similarly with the Gibbs adsorption energies but are generally several tenths of an eV higher. This means LGER applications using internal energies give reversible potential predictions that are off-set but will trend like those based on the Gibbs adsorption energies.

The adsorption energies appearing in Fig. 11.5, $U_{eff}$, and calculated activation energies for $O_2$(ads) dissociation are in Table 11.1. $U_{eff}$ are larger for $O_2$ and closer to 1.23 V. The activation Gibbs energies, $D_a^0$, for $O_2$ dissociation[7] increase as $U_{eff}(O_2)$ increases which may exclude this pathway for the surfaces. Reference [7] is an advanced DFT work that takes into account the effects of coverage on the kinetics and in it OOH(ads) dissociation was assumed to be fast and dissociation energies were not calculated for it. The activation energy for OOH(ads) formation increased slowly with potential in Ref. [7]. This explains why, though the onset potential for OOH(ads) is low at about 0.3 V it continues to form at higher potentials. The $U_{eff}$(OOH) values have small changes over the various skins, consistent with the slight effects seen

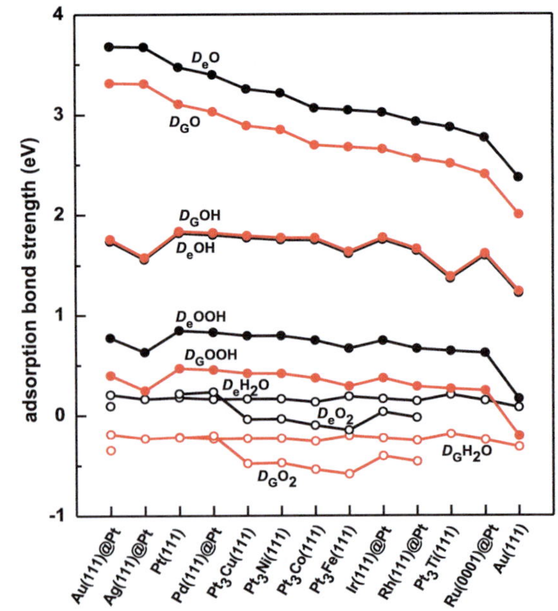


**Fig. 11.5** $D_e$ are calculated internal energies of adsorption for the molecules shown and $D_G$ are calculated Gibbs adsorption energies. Reactions and energies are tabulated in Table 11.1. Reprinted with permission from A. B. Anderson, R. Jinnouchi, and J. Uddin, "Effective reversible Potentials and Onset Potentials for $O_2$ Electroreduction on Transition Metal Electrodes: Theoretical Analysis," J. Phys. Chem. C 117, 41–48, 2013. Copyright 2013 American Chemical Society

**Table 11.1** Data used to make Fig. 11.5. Adsorption Gibbs bond strengths, $D_GO$ (eV) for O(aq); $D_GOH$ (eV) for OH(aq); $D_GOOH$(eV) for OOH(aq); $D_GO_2$(eV) for $O_2$(g); $D_GH_2O$(eV) for $H_2O$(l). $U_{eff}$(OOH)(V) for OOH(ads) dissociation pathway and $U_{eff}(O_2)$(V) for $O_2$ dissociation pathway. Activation Gibbs energy, $D_a^0$ (eV) for $O_2$ dissociative adsorption in the low-coverage limit. Reprinted with permission from A. B. Anderson, R. Jinnouchi, and J. Uddin, "Effective reversible Potentials and Onset Potentials for $O_2$ Electroreduction on Transition Metal Electrodes: Theoretical Analysis," J. Phys. Chem. C 117, 41–48, 2013. Copyright 2013 American Chemical Society

| Metal | *DGO* | *DGOH* | *DGOOH* | *DGO2* | *DGH2O* | *U*eff(OOH) | *U*eff(O2) | $D_a^0$ |
|---|---|---|---|---|---|---|---|---|
| Au(111)@Pt | 3.31 | 1.76 | 0.40 | –0.35 | –0.19 | 0.77 | 0.85 | 0.45 |
| Ag(111)@Pt | 3.31 | 1.56 | 0.25 | – | –0.23 | 0.78 | 0.85 | 0.45 |
| Pt(111) | 3.10 | 1.84 | 0.47 | –0.22 | –0.22 | 0.82 | 0.96 | 0.54 |
| Pd(111)@Pt | 3.03 | 1.82 | 0.45 | –0.21 | –0.23 | 0.84 | 1.00 | 0.58 |
| Pt3Cu(111) | 2.89 | 1.79 | 0.42 | –0.48 | –0.23 | 0.88 | 1.06 | 0.79 |
| Pt3Ni(111) | 2.85 | 1.77 | 0.42 | –0.48 | –0.23 | 0.89 | 1.08 | 0.96 |
| Pt3Co(111) | 2.68 | 1.77 | 0.37 | –0.54 | –0.26 | 0.92 | 1.16 | 1.10 |
| Pt3Fe(111) | 2.67 | 1.63 | 0.29 | –0.59 | –0.21 | 0.94 | 1.17 | 1.08 |
| Ir(111)@Pt | 2.65 | 1.77 | 0.37 | –0.41 | –0.23 | 0.93 | 1.18 | 1.12 |
| Rh(111)@Pt | 2.56 | 1.66 | 0.29 | –0.46 | –0.25 | 0.96 | – | 1.24 |
| Pt3Ti(111) | 2.50 | 1.38 | 0.26 | – | –0.19 | 1.04 | – | 1.26 |
| Ru(0001)@Pt | 2.40 | 1.61 | 0.24 | – | –0.25 | 1.00 | – | 1.43 |
| Au(111) | 2.00 | 1.23 | –0.21 | – | –0.31 | 1.08 | – | – |

experimentally, where improvements of a few tens of millivolts generally are the best that have been achieved. Reference [7] includes a volcano plot showing $Pt_3Ni$ is most active. From Table 11.1 this has a $U_{eff}$ value of 0.98 V. Entries with higher and $U_{eff}$ values show lower activity on the volcano plot. The lower activity was attributed in Ref. [7] to difficulty in forming OOH the first step reaction intermediates. Surfaces with $U_{eff}$ smaller than 0.98 V also have lower activity on the volcano plot which was attributed the difficulty in removing the more strongly bonded intermediates. Such behavior is an example of the Sabatier Principle. The implication for interpretations made using $U_{eff}$ values is that the ideal value is a limit less than 1.2 V due to the Sabatier Principle.

Others have calculated activation internal energies for $O_2$ dissociation when adsorbed on Pt(111). There are differences in the functional used for the DFT calculations and surface conditions in addition to the fact that the energies are internal rather than Gibbs energies used in Table 11.1. In one study the calculated activation energy was 0.69 eV[8] which is under the circumstances comparable to 0.54 eV in Table 11.1. In another 0.77 eV was calculated [9]. This reference also calculated an activation energy of 0.23 eV for dissociating OOH(ads) to O(ads + OH(ads) on Pt(111), a low value as expected.

## 11.7 More on Activation Energies and on Descriptors During $O_2$ Reduction

Explaining the limiting potential for $O_2$ reduction on platinum surfaces has not been a goal in many theoretical studies. Rather the goal has been calculating activating energies and relating them to kinetics of reaction [10]. The calculated reversible potentials for the electron transfer steps could sometimes have been deduced from the calculated data using LGER but the value of doing so was not recognized. Attempts to make purely kinetic explanations for the electrocatalysis by focusing on, for example OH(ads) removal, can miss the thermodynamics of the electrochemistry which establish the current onset potential and potential of limiting maximum current in Fig. 11.2. A descriptor for $O_2$ reduction is often the energy of the OH bond to the catalyst surface, the goal being to develop catalysts that reduce the strength of the bond. A variation on this was introduced by Luo and Koper [11]. They proposed the onset potentials for the O(ads) reduction to OH(ads) seen in cyclic voltammograms in pure electrolytes, absent $O_2$, as a descriptor. The onset potentials were close to 1.0 V. The polarization curves they measured were shaped similar to Fig. 11.2. This onset potential is close to the calculated $U_{\text{eff}}$ in Table 11.1 and to the OH(ads) and O(ads) reduction potentials discussed above. So the O(ads) to OH(ads) reduction is the highest potential step and can be considered a descriptor. However, $U_{\text{eff}}$ can be viewed as an overall controlling parameter.

## 11.8 Addressing a Complex Problem with DFT: Mechanism of Electroless Copper Deposition from pzc Calculations

Electroless deposition of metals is not an ordinary plating process taking place in a half-cell where metal cations are reduced to metal atoms at the cathode and deposited. It is a process for metal deposition on any surface, metallic or insulating[12]. The deposition is created by reduction of metal cations in solution by soluble organic reducing agents. Deposition is favored to occur on surfaces that can catalyze the reduction reaction [13]. Determining mechanisms for surface catalyzed electroless deposition continues as a challenge for experiment and theory. This section outlines a combined experimental and theoretical study from several years ago of electroless deposition of copper [14]. The work demonstrates how advanced theory can approach complex questions, but the need to adopt models for structure and composition leaves the conclusions in need of future confirmation. Applications of newer theories and better models along with further experimental characterization of this industrially important process are expected in the future.

The study employed the Interface code, including MPB for calculating reversible potentials at the pzc and the Gaussian code for finding C-H bond dissociation activation energies. Either formaldehyde, $H_2CO$ or glyoxylic acid, HC(O)COOH, which both contain the aldehyde group, RHCO are effective reducing agents and deprotonated ethylenediamine tetraacetic ($EDTA^{4-}$) is a complexing agent for $Cu^{2+}$ in aqueous solution. The overall reduction reaction,

$$Cu^{II}EDTA^{2-}(aq) + 2e^{-} \rightarrow Cu + EDTA^{4-}(aq) \tag{11.6}$$

has a standard reversible potential −0.213 V [15]. The aldehydes are hydrolyzed in acid to become diols, $RCH(OH)_2$ and then when introduced into the alkaline electroless deposition medium they become anions $RCH(OH)O^{-}$ which is their state as reducing molecules. The reactions causing reduction are

$$H_2C(OH)O^{-}(aq) \rightarrow HCOOH(aq) + 1/2H_2(g) + e^{-}(E_f) \tag{11.7}$$

and

$$\left(^{-}OOC\right)(H)C(OH)O^{-}(aq) \rightarrow \left(^{-}OOC\right)COOH(aq) + {}^{1}\!/_{2}H_2(g) + e^{-}(E_f) \tag{11.8}$$

After adjusting the electron energies, ($E_f$), to achieve equilibrium for these reactions, the respective calculated reversible potentials are –1.272 and –1.217 V, which are practically the same values, so that either reaction can reduce $Cu^{II}EDTA^{2-}(aq)$. The reductant methane diol in alkaline solution, $H_2C(OH)O^{-}(aq)$ was employed in the calculations because of its smaller number of atoms.

Reducing $Cu^{II}$ to $Cu^{0}$ as a two-electron process and in all likelihood goes in two steps passing through a short-lived $Cu^{I}$ intermediate. In the case of Eq. (11.8), one mole of the diol anion produces one mole of formic acid along with a half mole on hydrogen gas and one mole of electrons. It takes two moles of the diol anion to make one mole of $H_2(g)$ and the two electrons needed for reducing $Cu^{II}$ to $Cu^{0}$. The diol anions should repel one another and so each oxidation should form H atoms in some state and these atoms combine by a mechanism proposed below. Making the H atoms means breaking a strong C-H bond and so the H atoms must be stabilized by bonding to a molecule or surface before forming $H_2(g)$. Experiments have shown glyoxylic acid oxidation in base is faster over rough electroless copper surfaces than over smooth ones [16]. This suggests the state of the H atoms is bonded to sites in the copper surface prior to forming $H_2$ by combination. Low-coordinate atoms in surfaces are expected to form stronger interactions with adsorbates. This expectation was confirmed with calculations on cluster models using the Gaussian code incorporating the polarized continuum model to approximate hydration stabilizations. A lower barrier for H abstraction was predicted from the methane diol anion over a Cu adatom compared to over Cu as a member of the smooth (111) surface. This is shown in Fig. 11.6, where the calculated activation energies are 0.257 eV over the model of the smooth surface and 0.0608 eV over the adatom. A formic acid or oxalic acid anion form as the C-H bond breaks, depositing H on a Cu site. Calculations with the interface code show the complexes are strongly reducing, with respective reversible potentials –2.267 and –1.032 V. Both will reduce $Cu^{II}EDTA^{2-}(aq)$ to give $Cu^{0}$ atoms

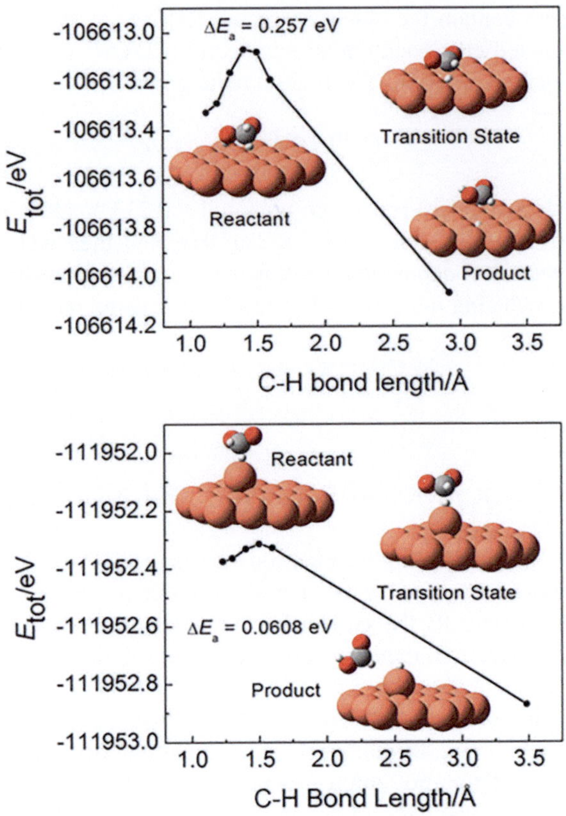


**Fig. 11.6** Top panel shows the methane diol anion adsorbed on the Cu(111) cluster model followed by a C-H bond stretching to a transition state and leaving H and an electron on the surface while formic acid departs. The lower panel shows the reaction when H is deposited on a Cu adatom. Reprinted with permission from M. Zhao, L. Yu, R. Akolkar, and A. B. Anderson, "Mechanism of Electroless Copper Deposition from $[Cu^{II}EDTA]^{2-}$ Complexes Using Aldehyde-based Reductants," J. Phys. Chem. C 120, 24,789–24,793, 2016. Copyright 2016 American Chemical Society

by Eq. (11.6) with its –0.213 V reversible potential. This means the electron that is released by the H abstraction will not stay with the acid molecule that is forming.

Cupric ion reduction can proceed in two steps:

$$Cu^{2+}(aq) + e^- \rightarrow Cu^+(aq) \quad (11.9)$$

followed by

$$Cu^+(aq) + e^- \rightarrow Cu^0 \quad (11.10)$$

The respective standard reversible potentials are 0.15 and 0.52 V. The reduction potentials are expected to be different for the EDTA complexes. For the reaction

$$Cu^{II}EDTA^{2-}(aq) + e^- \rightarrow Cu^{I}EDTA^{3-}(aq) \quad (11.11)$$

followed by

$$Cu^{I}EDTA^{2-}(aq) + e^{-} \rightarrow Cu^{0} + EDTA^{4-}(aq) \tag{11.12}$$

the calculated reversible potentials are, respectively, –0.571 and 0.241 V. Therefor the calculated value for the two-electron process is –0.165 V which is close to the –0.213 V measured value. This agreement is not unusual for the theoretical model given that, as shown in Chap. 9, the average error in predictions was 0.19 V for 34 small molecules whose reversible potentials covered over a six-volt range.

What are the structures of the copper EDTA complexes and how are the Cu atoms released? As the C-H bond is broken in the reducing molecules an energetic electron is released and will find its way to a nearby EDTA complex or to the Cu surface where it should have a long mean free path length based on the bulk value of 39.9 nm at room temperature [17]. This gives it a chance to encounter a complex before losing its potential energy thermally as phonons. However, even if the electron thermally equilibrates with the surface it should still able to carry out either of the reduction steps. This is because the pzc of Cu(111) surfaces is around –0.7 V in alkaline electrolytes[18] and because both complexes have a reduction potential greater than –0.7 V, the excess electron at the top of the metal valence band reduces them.

Calculations show the $Cu^{II}$ complex has octahedral coordination by six O atoms and upon the first reduction the complex begins to open up so that the coordination of $Cu^{I}$ is reduced to fourfold. With the final reduction to $Cu^{0}$ the coordination is twofold and the Cu atom is exposed. Figure 11.7 shows these structures.

By considering all the information obtained to this point a mechanism for electroless copper deposition can be proposed. Figure 11.8 puts it all together for methyl diol as the reducing agent.

The $Cu^{II}EDTA^{2-}(aq)$ complex regenerates since the bath contains copper sulfate and excess $EDTA^{4-}$ and the formation constant for the complex is high, with $\log K_f = 18.8$ [15].

A foundational idea in electrochemistry is the mixed potential theory [19]. Applied to electroless deposition in the present context, the theory postulates the oxidation of the reducing agent and reduction of metal complex are independent half-reactions taking place independently at different sites on the surface and when the currents are

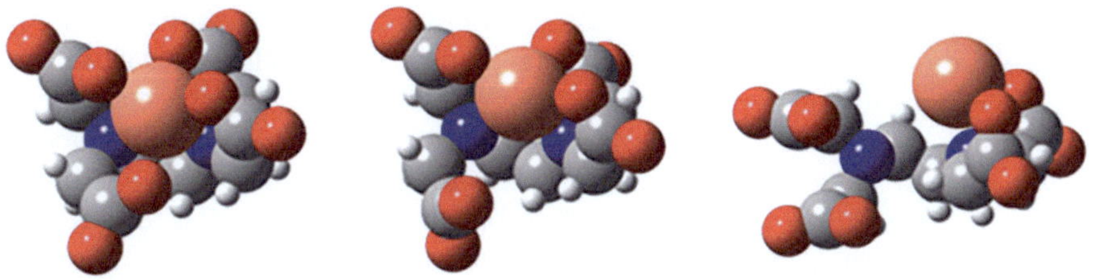

**Fig. 11.7** Three structure pictures show the Cu EDTA complex opening up as the Cu oxidation state decreases going from $Cu^{II}$ to $Cu^{I}$ to $Cu^{0}$. Reprinted with permission from M. Zhao, L. Yu, R. Akolkar, and A. B. Anderson, "Mechanism of Electroless Copper Deposition from $[Cu^{II}EDTA]^{2-}$ Complexes Using Aldehyde-based Reductants," J. Phys. Chem. C 120, 24,789–24,793, 2016. Copyright 2016 American Chemical Society

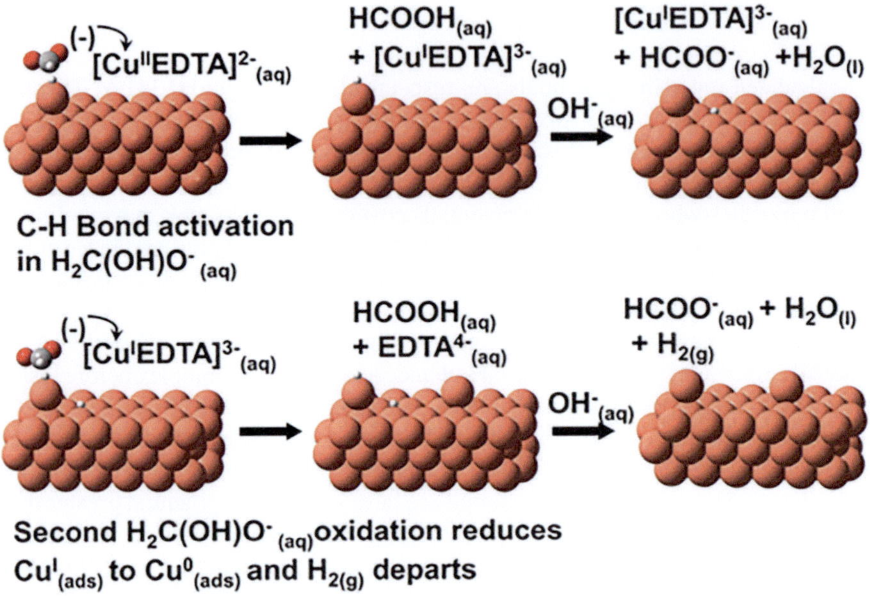


**Fig. 11.8** Proposed electroless copper deposition mechanism. In the top three pictures the diol transfers H to an adatom and the electron transfers to the $Cu^{II}$ complex. The result is formic acid + the $Cu^{I}$ complex. This is followed by proton abstraction from the formic acid by $OH^-$, generating formate. The bottom three pictures show the process repeated so that $Cu^{I}$ becomes $Cu^{0}$ with the Cu atom and a second H transferred to the surface followed by $H_2$ formation and desorption. Reprinted with permission from M. Zhao, L. Yu, R. Akolkar, and A. B. Anderson, "Mechanism of Electroless Copper Deposition from $[Cu^{II}EDTA]^{2-}$ Complexes Using Aldehyde-based Reductants," J. Phys. Chem. C 120, 24,789–24,793, 2016. Copyright 2016 American Chemical Society

balanced the surface is at a mixed potential. To test the postulate, current measurements were made using an electrolyte containing 0.036 M copper sulfate with 0.24 M $EDTA^{4-}$ complexing anions, and 0.19 M glyoxylic acid with pH adjusted to 12.5. Using a smooth Cu rotating disk electrode, current density varied with potential as shown by the red line in Fig. 11.9. Current density is zero at –0.43 V, meaning the external circuit is not contributing to oxidation or reduction reactions on the electrode surface. From this number and from this line alone there is no telling what the electroless deposition rate is at the measured mixed potential, but weight gain measurements showed it is over six times greater than at the theoretical −0.29 V mixed potential. The theoretical value was determined by measuring current densities with smooth copper electrodes for separate baths and the blue curve in Fig. 11.9 is measured for the glyoxylic acid bath and the black curve is measured for the $Cu^{II}EDTA^{2-}$(aq) bath. The currents balance at the theoretical mixed potential, which occurs at −0.29 V. In the complete electroless bath, at potentials positive of the measured −0.43 V mixed potential, some of the glyoxylic acid is being oxidized, sending electrons to an external half-cell while electroless deposition is taking place and for potentials

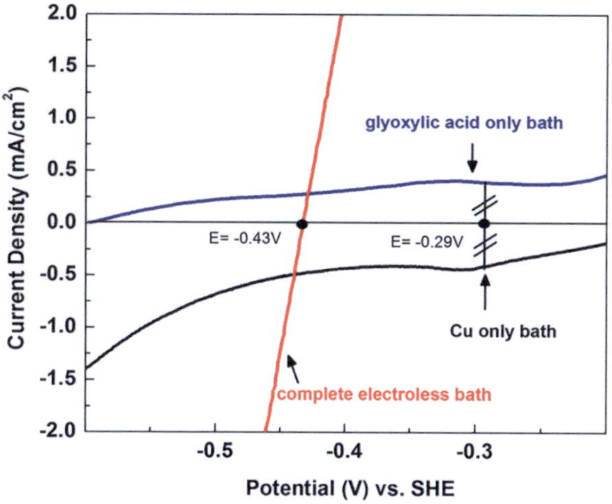


**Fig. 11.9** Polarization curves obtained on a Cu RDE (200 rpm) from a complete electroless bath, red line, a glyoxylic acid bath, blue curve, and a $Cu^{II}EDTA^{2-}$(aq) bath, black curve. A discrepancy is evident between −0.29 V predicted by the mixed potential theory and −0.43 V measured experimentally in the complete electroless bath. Reprinted with permission from M. Zhao, L. Yu, R. Akolkar, and A. B. Anderson, "Mechanism of Electroless Copper Deposition from $[Cu^{II}EDTA]^{2-}$ Complexes Using Aldehyde-based Reductants," J. Phys. Chem. C 120, 24,789–24,793, 2016. Copyright 2016 American Chemical Society

negative of −0.43 V some of the $Cu^{II}EDTA^{2-}$(aq) is reduced, accepting electrons from an external half-cell during electroless deposition.

The model for the theoretical mixed potential envisions independent half-cell reactions taking place on the electrode surface and connected by electron transport through the conduction band as in Chap. 6. A steady state is reached when the donor level of the species being oxidized matches the acceptor level of the species being reduced. However, the theoretical and experimental findings here indicate the half-reactions are not isolated and independent. A recent paper measures weight gains at a series of potentials applied to an electroless bath to provide further perspective concerning theoretical mixed potentials and measured ones [20]. Electroless deposition may be roughening the initially smooth electrode surface, activating it for the oxidation of the oxalic acid anion. The oxidation step has a strong thermodynamic driving force and it seems the low-coordinate Cu sites in the surface provide a low activation energy pathway for generating the electrons that reduce $Cu^{II}EDTA^{2-}$(aq). This is a catalytic effect allowing faster reaction compared to isolated oxidation and reduction reactions on the surface. More work needs to be done to fully explain the electron transfer process. Interestingly, attempts to develop catalyst theories based on the mixed potential concept are a recent development [21].

In conclusion, theory provides a plausible mechanism and understanding of electroless copper plating in alkaline baths containing $Cu^{II}EDTA^{2-}$(aq) with methyl diol. Experiments find the measured mixed potential when glyoxylic acid is used for the oxidant is different from the theoretical value determined form the potential of equal but opposite sign current densities when measured for separate baths of the two components. This means the two half-reactions are influencing one another and new research is needed to fully clarify the mechanism.

## References


1. Bard AJ, Parsons R, Jordan J (eds) (1985) Standard aqueous potentials in aqueous solution. Marcel Dekker, New York
2. Stanbury DM (1989) Reduction potentials involving inorganic free radicals in aqueous solution. Inorg Chem 33:69–138
3. Lide DR (ed) (2005) CRC handbook of chemistry and physics, 86th edn. CRC, Boca Raton
4. Anderson AB (2012) Insights into electrocatalysis. Phys Chem Chem Phys 14:1330–1338
5. Tian F, Anderson AB (2011) Effective reversible potential, energy loss, and overpotential on platinum fuel cell cathodes. J Phys Chem C 115:4076–4088
6. Anderson AB, Jinnouchi R, Uddin J (2013) Effective reversible potentials and onset potentials for $O_2$ electroreduction on transition metal electrodes: theoretical analysis. J Phys Chem C 117:41–48
7. Jinnouchi R, Kodama K, Hatanaka T, Morimoto Y (2011) First principles based mean field model for oxygen reduction reaction. Phys Chem Chem Phys 13:21070–21083
8. Mäkinen M, Laasonen K (2023) Density functional study of trends in water dissociation on oxygen-preadsorbed and pure transition metal surfaces. Surf Sci 734:122305
9. Haile AS, Yohannes W, Mekonnen YS (2020) Oxygen reduction reaction on Pt-skin $Pt_3V(111)$ fuel cell cathode: a density functional theory study. RCS Adv 10:27346–27356
10. Sha Y, Yu TH, Merinov BV, Shirvanian P, Goddard WA (2011) Oxygen hydration mechanism for the oxygen reduction reaction at Pt and Pd fuel cell catalysts. Phys Chem Lett 2:572–576
11. Luo M, Koper MTM (2022) A kinetic descriptor for the electrolyte effect on the oxygen reduction kinetics on Pt(111). Nat Cal 5:615–623
12. Mallory GO, Hajdu JB (eds) (1990) Electroless plating: fundamentals and applications. American Electroplaters and Surface Finishers Society Inc., Orlando FL
13. Dubin VM, Shacham-Diamand Y, Zhao B, Vasudev PK, Ting CH (1997) Selective and blanket electroless copper deposition for ultralarge scale integration. J Electrochem Soc 144:898–908
14. Zhao M, Yu L, Akolkar R, Anderson AB (2016) Mechanism of electroless copper deposition from $[Cu^{II}EDTA]^{2-}$ complexes using aldehyde-based reductants. J Phys Chem C 120:24789–24793
15. Li J, Kohl PA (2004) Complex chemistry & the electroless copper plating process. Plat Surf Finish 91:40–46
16. Yu L, Guo L, Preisser R, Akolkar R (2013) Autocatalysis during electroless copper deposition using glyoxylic acid as reducing agent. J Electrochem Soc 160:D3004–D3008
17. Gall D (2016) Electron mean free path in elemental metals. J Appl Phys 119:085101-1-085101-5
18. Auer A, Ding X, Bandarenka AS, Kunze-Liebhäuser J (2021) The potential of zero charge and the electrochemical interface structure of Cu(111) in alkaline solutions. J Phy Chem C Nanomater Interfaces 125:5020–5028. https://doi.org/10.1021/acs.jpcc.0c09289
19. Wagner C, Traud W (1938) The interpretation of corrosion phenomena by super-imposition of electrochemical partial reactions and the formation of potentials of mixed electrodes. Z Elektrochem 44:391–402
20. Zeszut R, Landau U (2019) Determination of redox currents in electroless systems: correct application of mixed potential analysis. J Electrochem Soc 166:D737–D741
21. Yan M, Namari NAP, Nakamura J, Takeyasu K (2024) Theoretical framework for mixed-potential-driven catalysis. Commun Chem 7(69):1–7

# Chapter 12
# Recent Theoretical Studies of pH Dependence of Several Surface Reactions

**Abstract** A level of theoretical modeling has been reached in the Interface and similar codes that makes it possible to explore new measurable trends in electrochemistry that depend on potential-dependent variations in surface charge and bond polarizations. Though there are approximations in the modeling giving uncertainties yet to be refined in the calculated energies, this chapter illustrates various applications where the predictions at least qualitatively agree with measured phenomena. Disagreements with measured trends provide goals for future improvement in the theoretical modeling. Included are studies of (i) pH dependence of surface reactions without surface charging, (ii) calculations of acid constants used to derive the Nernst equation, (iii) pH dependence of surface reactions with surface charging, (iv) surface structure of water overlayers as functions of applied potential, (v) width of the potential range defining the double layer on Pt(111) electrodes, (vi) the importance of hydrogen bonding on adsorbed OH, and (vii) Pourbaix diagrams for reacting surfaces.

## 12.1 General Introduction

Throughout earlier chapters of this book, except for Chaps. 6 and 7 which characterize electronic structures a recurring criticism applies that the electrochemical interface computations at the potential of zero charge (pzc) omit how changes in electrode potential affect bond polarizations and strengths. Up until the development of Interface and related codes incorporating surface charging and modified Poisson-Boltzmann (MPB) theory to approximate electrolyte response to gain the desired electrode potential, most theoretical studies of electron transfer reactions at electrode surfaces were performed at the pzc. Nevertheless, they provided a widely accepted foundation for understanding electrocatalytic reactions. In this chapter it is shown the pzc can change by several tenths of a volt depending on the structure of an over layer on a Pt(111) electrode. Using interface with MPB, pH dependencies of reversible potentials for H deposition and OH(ads) formation from $H_2O$(ads) oxidation are calculated. The results are used to explain the variable double layer potential

 
A. Anderson, *Past and Current Models in Electrochemistry*,
https://doi.org/10.1007/978-3-031-97791-6_12

range and to suggest the possibility of deviations from Pourbaix diagrams when bulk equilibria are assumed to apply to surface equilibria.

## 12.2 Calculating pH Dependence of Surface Oxidation and Reduction Reactions Using pzc Models

Acidity of electrolytes, usually expressed as pH, which is the negative of the $\log_{10}$ of the hydrogen ion activity. The pH is a parameter widely used for characterizing and understanding biochemical and electrocatalytic reactions and equilibria. Incorporating pH into quantum theoretic calculations of equilibrium potentials and activation energies is a necessary component for assembling a comprehensive reaction theory. It was mentioned earlier how the Nernst equation was used to define the pH dependence of the electrode potential and Fermi energy of the reacting electron [1]. This was done for reactions with proton reactants or products actually calculated at the pzc and the potential referenced to the standard hydrogen electrode.

There are several steps to explaining why the reversible potential of a reaction depends on pH. Beginning with the equation for reducing an oxidized molecule, Ox, using $H^+$(aq) and an electron to form H-Ox, which is called Red for short,

$$\mathrm{Ox} + \mathrm{H}^+(\mathrm{aq}) + \mathrm{e}^-(U) \rightleftharpoons \mathrm{Red} \tag{12.1}$$

the Nernst equation for the reversible potential at non-standard conditions is

$$U_{\mathrm{rev}} = U^0 - RT/F\ln\left([\mathrm{Red}]/\{[\mathrm{Ox}]\left[\mathrm{H}^+(\mathrm{aq})\right]\right)\} \tag{12.2}$$

These two equations apply to solution equilibria. When one or both of Ox and Red are adsorbed on the electrode, their potential-dependent adsorption energies will enter into the energy balance of Eq. (12.1). Consequences of the potential dependencies of adsorption of reactants and products come later in the discussion of Pourbaix diagrams. When Eq. (12.1) is the hydrogen evolution reaction,

$$\mathrm{H}^+(\mathrm{aq}) + \mathrm{e}^-(U) \rightleftharpoons {}^1\!/_2\,\mathrm{H}_2, \tag{12.3}$$

the Nernst equation is written

$$U_{\mathrm{rev}} = U^0 - 2.3026[RT/F]\mathrm{pH} - [RT/2F]\ln p(\mathrm{H}_2)/p^0 \tag{12.4}$$

Potentials in Eqs. (12.2) and (12.4) are referenced to a standard hydrogen half-cell and the working electrode is a half-cell with adjustable pH. The half-cells are typically connected by an electrolyte salt bridge. Temperature-dependent errors in potential measurements are known to be created by salt bridge interfaces with the electrolyte solutions [2]. However, corrections for this effect are small enough that they are not applied in most work. The standard hydrogen half-cell has $H_2$ gas at pressure $p^0$ = 1.0 bar bubbled over a platinum electrode immersed in standard electrolyte with $H^+$(aq) activity 1.0 M. As is often done, concentration notation is used in the following to represent activities, though activities are products of activity coefficients times the concentrations. The activity coefficients appear in the thermodynamic derivation of equations such as Eqs. (12.2) and (12.4). Activity coefficients are close to unity for solution components of low concentration. The half-cell with the solution whose pH is being measured contains a platinum electrode and the potential $U$ is the voltage measured between the electrodes in the two cells. The 2.3026 $RT/F$ factor, which equals 0.059160 V at 298.15 K, causes the potential $U$ to go from 0.0 V, which is the SHE potential at pH = 0 to −0.828 V at pH = 14. This also means the electron chemical potential and $E_f$ of the electrode immersed in the solution of adjustable pH are destabilized by 0.828 eV at pH = 14 relative the value of the standard electrode with pH = 0. This was depicted in Chap. 7 when considering properties of H-terminated diamond surfaces in the ambient atmosphere. All four reaction steps combining an electron and proton in the four-electron reduction of $O_2$ to water experience 0.828 V decrease in reversible potential when the pH is increased from 0 to 14. The change is linear with pH for these reactions but in other cases the rate may depend on reaction stoichiometry. In the Zhang-Anderson paper that was discussed in detail in Chap. 8, all calculations of activation energies using local reaction center (LRC) model applied for reduction at pH 13 by shifting the potential scale negative by 0.770 V [1]. This means the external electron source is destabilized by 0.770 eV from its value for pH = 0. $H_2O$ and not $H^+$(aq) was used for the reaction in this case because at this pH there are too few $H^+$(aq) to make significant contribution to measured current so $H_2O$ is the proton source. At pH = 1 the potential scale is shifted 0.0592 V negative and hydronium ions are the reactant [3]. Potential scale shifts of this type apply to the LRC calculations but such calculations take place at the pzc with the electron coming from or going to an external source or sink.

A theoretical study of this type, excluding the double layer and surface charging, was of pH dependencies of reaction rates for steps during CO reduction on copper [4] Solution and intermediate surface concentrations became parameters in the model and current densities and Tafel slopes were calculated. The slow step was found to be coupling of two CO(ads) to form OC-CO(ads). Supplemental material to the article discussed sources of errors in the model and, notably, for the adsorbed intermediates the work functions had varying values. Determining resulting errors from the lack of potential control was beyond the scope of the work.

Calculations with potential control are discussed next, including structures and properties for adsorbates and reversible potentials for reactions. Though CO reduction on Cu electrodes is not addressed, the electrochemical vibrational tuning rate of CO/Pt is, followed by H deposition and $H_2O$ oxidation.

## 12.3 Applying First Principles Calculations with the Nernst Equation Requires Ability to Calculate an Accurate Water Constant

A first principles Nernst equation may be obtained based on the calculated value of the water equilibrium constant, $K_w$, which is the equilibrium constant for the reaction

$$H_2O(l) \rightleftharpoons H^+(aq) + OH^-(aq) \tag{12.5}$$

To do this, Zhao and Anderson calculated the Gibbs reaction energy, $\Delta_w G$, using the interface code, getting 0.845 eV at standard temperature, 298.15 K [5]. In the calculations the MPB theory and dielectric screening were employed and $H^+(aq)$ was modeled by $H_3O^+(H_2O)_3$ and $OH^-(H_2O)_3$ was used for $OH^-$. The standard state for $H_2O(aq)$ was modeled by an $(H_2O)_4$ ring with the $H_2O$ molecules connected by hydrogen bonds so that the energy of a single water molecule is ¼ the ring energy. Structures for the species were optimized by energy minimization using the variational theorem. Using the calculated $\Delta_w G$, the water equilibrium constant for Eq. (12.5),

$$K_w = \exp(-\Delta_w G/kT), \tag{12.6}$$

is $5.212(10^{-15})$. The $\log(K_w)$, $pK_w$, is then 14.283, compared to the experimental value 14.00. Alternative calculations of acidity constants using DFT and molecular dynamics have been reviewed [6]. Given this calculated $K_w$, in the Zhao approach $[RT/F]$ was replaced with 0.0604 so that the pH range remains 14 but the energy space is slightly distorted. Doing this scales the energy dependence on pH so that the potential decreases by 0.828 V from pH = 0 to pH = 14 and the electrode Fermi level goes from −4.454 eV at pH = 0 to −3.626 at pH = 14 while supporting a reaction of the form of Eq. (12.1) in equilibrium. This scaling makes no significant difference to results of calculations discussed below. Relationships of the Fermi energy ($E_f$), which is the top of the electrode valence band and equals the electrochemical potential of electrons on the electrode surface (Chap. 6), and potential with pH are shown in Fig. 12.1. This figure is significant because it shows how $E_f$ of an electrode participating in an equilibrium of the form of Eq. (12.1) changes when referenced to a standard electrode as the electrolyte pH is changed. Might this shift cause a shift in vibrational adsorption energies for molecules such as CO adsorbed on an electrode surface when the pH is changed? And how might surface coverage of reduction products affect pH dependence of reversible potentials?

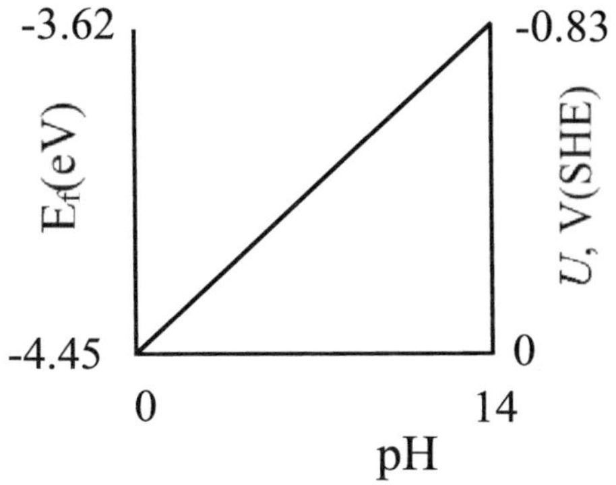


**Fig. 12.1** Calculated dependencies on pH of an electrode $E_f$ and potential $U$ for which a reaction of the form Eq. (12.1) is in equilibrium when $U$ is referenced to the standard hydrogen electrode. As pH goes from 0 to 14, $E_f$ goes from −4.45 eV to −3.62 eV and $U$ goes from 0 V to −0.83 V

## 12.4 Understanding pH Dependence of CO/Pt(111) Electrochemical Vibrational Tuning Rate (EVTR)

Results from experiment and theory discussed Chap. 5 show CO(ads) $d\upsilon/dU$ decreases as $\theta$ increases on Pt(111) [7]. Theory calculated that for 2/3 ML CO coverage on Pt(111) the tuning rate decreased as the potential increased by about 6 $cm^{-1}$ per volt but at the same time the electrode potential increased as the pH decreased. The corresponding calculated rate of $d\upsilon$/dpH was 1/3 the value that was estimated from plots in one experimental study [8] and 1/5 the rate estimated from plots in another [9]. Based on what is known so far, the implication of the slow calculated change as a function of pH is that if the coverage of CO were to decrease as the pH increases then that would contribute to increasing $d\upsilon$/dpH from the calculated value. However, the work of Uddin shows the adsorption bond strength of CO increases as the potential is shifted in the negative direction with increasing pH, which makes CO(ads) displacement seem unlikely [7]. From this it is concluded that there seem to be more dimensions to understanding the pH dependence of the vibrational tuning rate for CO adsorbed on Pt(111) electrodes than have been identified so far.

## 12.5 Formula for Determining pH Dependencies of Surface Reactions When Reactant and Product Adsorption Energies Vary with Electrode Potential: Forming upd H

Appealing again to the perturbation molecular orbital (PMO) theory of Chap. 5, pH-dependent shifts in $E_f$ and surface valence band orbital energies should have measurable effects on some surface reactions. For adsorbed Ox($U$) and Red($U$), consider the pH dependence of the surface reaction where protons are consumed:

$$\mathrm{Ox}(U) + \mathrm{H}^+(\mathrm{aq}) + \mathrm{e}^-(U) \rightleftharpoons \mathrm{Red}(U) \tag{12.7}$$

Zhao and Anderson showed that at equilibrium the following holds [5]:

$$\begin{aligned} G[\mathrm{Red}(U)] =& \Delta_{\mathrm{react}}G(U) + G[\mathrm{Ox}(U)] + G\big[\mathrm{H}^+(\mathrm{aq})\big] \\ & - 4.454\ \mathrm{eV} - FU + 0.0604\ \mathrm{pH(eV)} \end{aligned} \tag{12.8}$$

The 0.0604 factor that appears in place of 0.0592 in the equation is the value calculated using $pK_w = 14.283$. This substitution made no substantial difference to conclusions in the study. The advancement of using Eq. (12.7) rather than pzc computational model of Eq. (12.1) is the potential dependence for structures and energies of the adsorbed reactants and products are included in the equilibrium. This is an advancement over the CHE/LGER model which does not allow reactant and product structures to change with changes in potential since the potential is defined only in terms of the energy of the electron participating in the reaction. Therefor the calculated energies are inaccurate for all potentials except the potential corresponding to $E_f$ of the optimized structures, which is at the pzc when these models are used. A further complication is that the pzcs for the reactant and product will generally have different values.

Reversible potentials calculated using Eq. (12.8) were compared with experimentally measured reversible values referenced to the reversible hydrogen electrode [5, 10]. Since reversible potentials, $U_{\mathrm{rev}}$, are defined by oxidized and reduced species having the same energy, the predicted values may be found by fitting a polynomial to $\Delta_{\mathrm{react}}G(U)$ from Eq. (12.8) and finding the roots where the polynomial equals zero. An equivalent method, done graphically, is demonstrated in Fig. 12.2.

Here the left panel locates the roots at intersections of the $G[\mathrm{Red}(U)]$ and $G[\mathrm{Ox}(U)] + G(\mathrm{e}^-)$ functions. The right panel shows predicted linear dependence of $U_{\mathrm{rev}}$ on pH when the reduction product is 1/12 ML H(ads) on Pt(111). The reactant energies at each potential depend on pH through the Nernst equation, and they are shifted accordingly in the left panel. Because the energies of the 1/12 ML H(ads) product depend only slightly on pH, the 1/12 ML H(ads) pH = 0 curve is chosen for determining the intersections. To understand the method, it must be remembered that the energies graphed in the left panel are the total energies of the translational

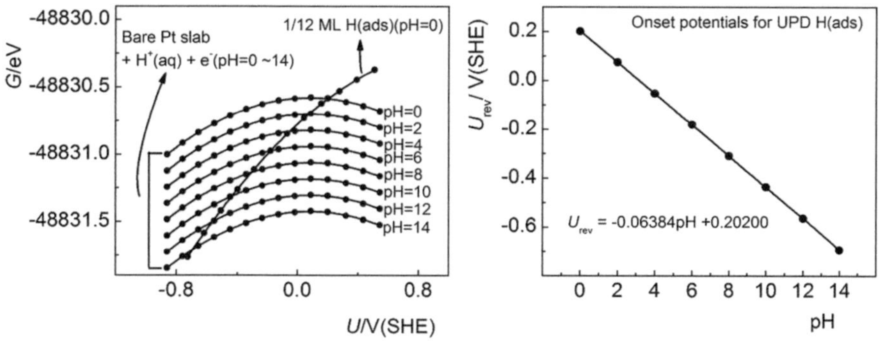


**Fig. 12.2** The first panel is a graphical determination of pH dependence of reversible potentials for forming 1/12 ML H(ads) on Pt(111). Total Gibbs energies, *G*, are energies of the reactants are on the vertical axis for pH values 0 to 14 in steps of 2 and a second curve representing the Gibbs energies of the products for all values of pH intersects at the calculated onset potentials. The second panel is a graph of the onset potentials showing their linear decrease with slope −0.06384 as pH increases. Reprinted with permission from M. Zhao and A. B. Anderson, "Prediction the Double Layer Width on Pt(111) in Acid and Base with Theory and Extracting it from Experimental Voltammograms," J. Phys. Chem. C 121, 28,051–28,064, 2017. Copyright 2017 American Chemical Society

unit cell modeling the electrode and the potential is changed by adding charge to the translational cell. The MPB theory employed in the calculations compensates the added surface charge and is parameterized for unit ionic strength, assuming ions of a standard size.

No water molecules were included in the H deposition calculations because water molecules interact weakly with H(ads) [11]. However, as will be illustrated below, water orientations vary with potential and were water molecules included, they would have some effect on the calculated energies. In this example the dependence of $U_{rev}$ on pH is linear with slope −0.06384, which is slightly different from −0.0604 for the theoretical Nernst equation. The difference in slope results from a weak dependence of the strength of the covalent bond between H and the surface on the band shifting at different potentials. As pointed out by Asiri and Anderson, the low-coverage potentials will increase approximately 0.15 V when the high configurational entropy at low coverage is taken into account [11]. This is done below when predictions are made of width of double layer potential range on Pt(111). Since the MPB procedure is parameterized using monovalent ions with unit ionic strength, this allows for $H^+$(aq) and $OH^-$(aq) to be viewed as contributing to the compensating electrolyte charge distribution along with any dissolved monovalent salt that might be used to maintain the unit ionic strength. However, any response of the hydronium ion at the surface to charging the surface is absent.

The model misses some structure and energy contributions. Experimental measurements on Pt(111) aqueous interfaces showed that interfacial capacitances at approximately constant pH near the pzc vary with changes in water molecule orientation at the surface, the size and hydration strength of the ions, and attractive ion-surface interactions [12]. These details go beyond the traditional Gouy-Chapman-Stern model with its assumptions of a polarizable diffuse layer above the Helmholtz double layer and it includes specifically adsorbed ions. Combined experimental and theoretical work on dilute aqueous sodium fluoride solutions over Ag(111) electrodes detailed similar interactions between the electrolytes and the electrodes, including specific adsorption of partially solvated fluoride anions [13]. From these two studies and others we see that details of double layers are becoming accessible experimentally and theoretically.

## 12.6 pH Dependence of Onset Potential for $H_2O$(ads) Oxidation to OH(ads)

The oxidation of $H_2O$(ads) to form OH(ads) is a case where the reaction onset potential is not a linear function of pH. As a prelude, determining the structure of water saturation at coverage on Pt(111) has been a goal of the surface science community. To this end, experimental procedures employed include low energy electron diffraction (LEED) [14], helium atom diffraction (HAD) [15], and temperature-variable scanning tunneling microscopy (STM) [16, 17] . From these studies it was deduced that water molecules form an ordered hydrogen-bonded hexagonal arrangement of on the surface. Doering and Madey proposed structures for water bilayers on Ru(001), which is isostructural to Pt(111), and one of them is labeled after them as the MD structure [18], which is one of the structures considered in the following.

A monolayer of H(ads) atoms interacts weakly with water molecules with little potential dependence. For a $H_2O$ molecular monolayer adsorbed on Pt(111), the oxygen atom forms weak lone-pair donation bonds to the surface and OH can also form weak hydrogen bonds to the surface at the pzc. But these bond strengths depend on the applied potential, which when changed can change the water structure. The result is the total energy calculated for the water oxidation reaction:

$$H_2O_{ads}(U) \rightleftharpoons H^+(aq) + e^- + OH_{ads}(U) \tag{12.9}$$

has nonlinear dependence on applied potential.

As the potential is increased through the double layer range, the onset potential for Eq. (12.9) on Pt(111) marks its high-potential end and the lower end is marked by the onset of hydrogen deposition. In the double layer potential range there are no surface reactions and the small current density measured in voltammograms for this potential region is caused by capacitive charging. The Interface code was used to calculate the onset potentials for Eq. (12.9) which along with the H(ads) onset potentials yielded estimates of the measured double layer widths [10]. In the study three hexagonal 2/3 ML water overlayer structures were optimized and a 1/6 ML OH(ads) + 1/2 ML $H_2O$(ads) structure representing low coverage of the OH(ads) water oxidation product was optimized too. Structures calculated for four electrode potentials are in Fig. 12.3.

Calculated onset potentials for OH(ads) formation as functions of pH for the three structures are shown in Fig. 12.4. As a function of pH, the MD structure has the highest slope and greatest deviation from linearity. Over the pH range 0 to 14, $U_{rev}$ increases by 0.580 V for H-down-2, by 0.751 V for H-down-1, and by 1.141 V for MD. Why do the rates and non-linearity vary dramatically with the structure of the interface for these cases but not for H(ads)? Because unlike the weakly interacting H(ads) surfaces, the OH(ads) surfaces interact relatively strongly with the $H_2O$ and this disrupts the structures of the water networks. However, the presence of H(ads) may also change the energy and structure of the hydrogen-bonded water molecule network at the interface and this is a topic calling for further exploration.

## 12.7 Width of Double Layer Potential Region on the Pt(111) Electrode

At a given value of pH, the difference between the onset potential for water oxidation to OH(ads) and the onset potential for $H^+$(aq) or $H_2O$(l) reduction to H(ads) is the width of the double layer. Double layer widths for different pH values were estimated from cyclic voltammogram data from the lab of Juan Feliu [19] and compared with calculated values. This was done by making spline fits to the voltammogram numerical data points and assigning the onset potentials to values with maximum second derivatives (Fig. 12.5). The onset potentials for the anodic and cathodic scans are nearly the same for pH = 1 as well as for graphs for the other pH values, reflecting the reversibility observed in the voltammograms. Double layer widths calculated for the three water structures on Pt(111) are plotted in Fig. 12.6 along with experimental estimates.

Comparing theoretical curves and experimental curves in Fig. 12.6, agreement is best for the H-down-1 structure. The point at pH 8.3 in the experimental curve is low and off-trend apparently due to the presence of bicarbonate in the electrolyte in this case. Bicarbonate probably stabilizes H(ads) or OH(ads) and these stabilizations increase $U_{rev}$ for H(ads) and decrease $U_{rev}$ for OH(ads), thereby reducing the width of the double layer.

Using a larger translational cell and more water molecules, calculated double layer widths for forming 1/12 ML OH(ads), when starting with the H-down-1 water structure, show closer average agreement with experiment but curvature of the calculated line is slightly negative, see Fig. 12.7. Estimated onset potentials using data from the Feliu lab for forming upd H(ads) vary as a function of pH over a small 0.04 V range on the RHE scale. Since the calculated double layer potential widths in Fig. 12.7 are close to experiment, the onset potentials calculated for OH(ads) formation are closer to the experimental measurements. This suggests the OH adsorbs more strongly at the lower coverage and this causes the calculated onset potentials to be less. The

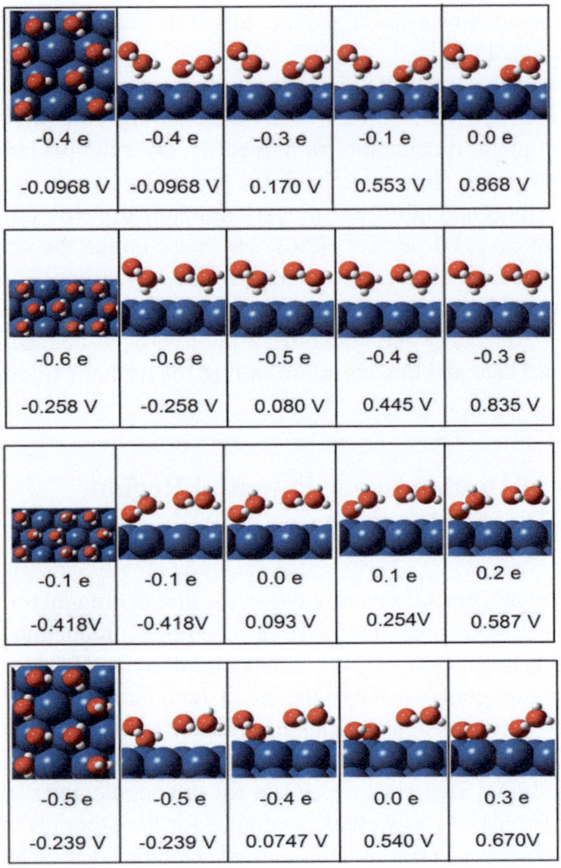


**Fig. 12.3** Platinum atoms are blue, oxygen atoms red, and hydrogen atoms white. Given potentials depend on added charges shown in units of electron charge. Top row: 2/3 ML $H_2O$(ads) H-down-1. Second row: 2/3 ML $H_2O$(ads) H-down-2. Third row: 2/3 ML $H_2O$(ads) MD structure. Bottom row: 1/6 ML OH(ads) + 1/2 ML $H_2O$(ads). Each structure has five panels, the first a top down view and the following four are cross-section views of the interface. The first shows hexagonal water structures and the next four panels show how an increase in potential causes increased lone-pair donation bonding so that the third $H_2O$ in the H-down-1 structure case to move closer to the metal. In the H-down-2 case there is no lone-pair bonding to the surface and the third $H_2O$ moves up as the potential increases. In the MD structure case the first $H_2O$ is lone-pair bonded to the surface and moves slightly closer at higher potentials. In the last row the first $H_2O$, which is hydrogen bonded to OH(ads) and then the third $H_2O$ drop to the surface as the potential increases to ~ 0.5 V and ~ 0.6 V, respectively. Reprinted with permission from M. Zhao and A. B. Anderson, "Prediction the Double Layer Width on Pt(111) in Acid and Base with Theory and Extracting it from Experimental Voltammograms," J. Phys. Chem. C 121, 28,051–28,064, 2017. Copyright 2017 American Chemical Society

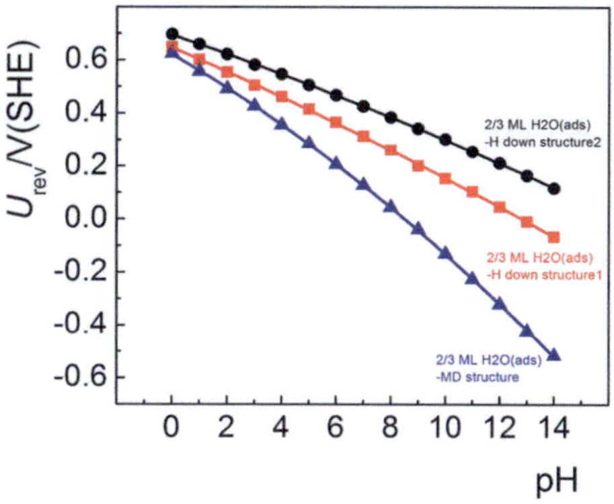


**Fig. 12.4** Reversible potentials on the vertical axis for 2/3 ML $H_2O$(ads) oxidation to 1/2 ML $H_2O$(ads) + 1/6 OH(ads) calculated for the three water structures over pH = 0 to pH = 14 range on the horizontal axis. Reprinted with permission from M. Zhao and A. B. Anderson, "Prediction the Double Layer Width on Pt(111) in Acid and Base with Theory and Extracting it from Experimental Voltammograms," J. Phys. Chem. C 121, 28,051–28,064, 2017. Copyright 2017 American Chemical Society

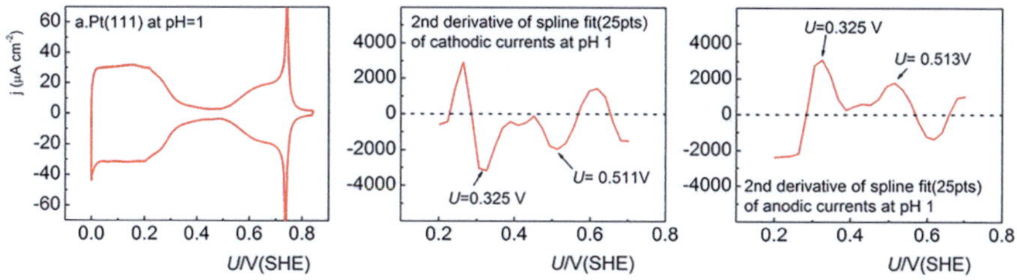


**Fig. 12.5** Left-hand panel shows the cyclic voltammogram from a Pt(111) electrode in 0.1 M $NaClO_4$. The next two panels graph the second derivatives of spline fits to the current densities of the positive and negative scans. Reaction onset potentials are assigned as the potentials where the second derivatives are zero. The first of these marks H(ads) and OH(ads) onset potentials at 0.325 V and 0.511 V for the cathodic branch and the second marks onset potentials at 0.325 V and 0.513 V for the anodic branch. Reprinted with permission from M. Zhao and A. B. Anderson, "Prediction the Double Layer Width on Pt(111) in Acid and Base with Theory and Extracting it from Experimental Voltammograms," J. Phys. Chem. C 121, 28,051–28,064, 2017. Copyright 2017 American Chemical Society

reason for the change in curvature is unknown and presents a challenge for future researchers to reproduce and explain.

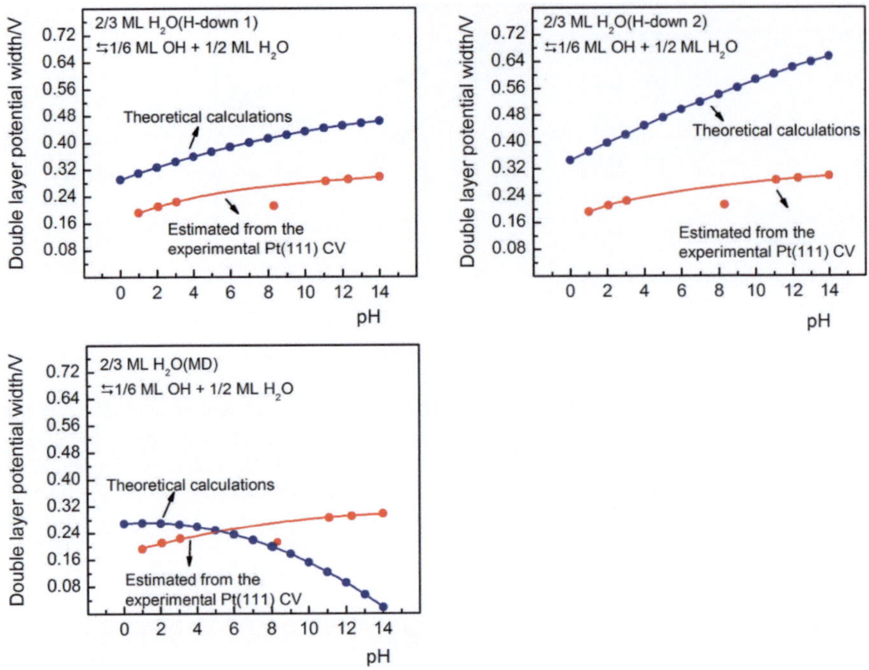


**Fig. 12.6** Calculated widths of the double layer potential region and experimental estimates over pH range 0 to 14. For clarity the $H^+(aq) + e^-$ contributions to the right-hand sides of the reactions are omitted in the figures. Of the three panels the H-down-1 curve best parallels the experimental curve, trending toward lower slope as pH goes from 0 to 14, though it overestimates the double layer width by more than 0.1 V. Reprinted with permission from M. Zhao and A. B. Anderson, "Prediction the Double Layer Width on Pt(111) in Acid and Base with Theory and Extracting it from Experimental Voltammograms," J. Phys. Chem. C 121, 28,051–28,064, 2017. Copyright 2017 American Chemical Society

## 12.8 Dependence of Pt(111) Voltammogram Shapes on pH for the Two Oxidation Waves Up to 1.2 V and the Importance of Hydrogen Bonding on Surface States of OH(ads)

The sharp current peak for the pH 1electrolyte in Fig. 12.5 shows that as the potential increases the final OH(ads) coverage forms over a narrow potential range. At pH = 0, the calculated onset potential 0.643 V for forming 1/2 ML OH(ads) (Table 12.1) is 0.643 V and a slightly smaller value 0.634 V is seen for forming 1/3 ML OH(ads). From the predictions in Table 12.2 it may be concluded that for the low pH values there are two dominating ranges of potentials, one around 0.6 V to 0.5 V for forming 0 ML to 1/3 ML OH(ads) and a second around 1.3 V to 1.0 V for forming 1/2 ML to 2/3 ML OH(ads). As pH increases to 14 the two groups merge to a single value around 0.0 V. The Alicante group estimated from integrations of measured current

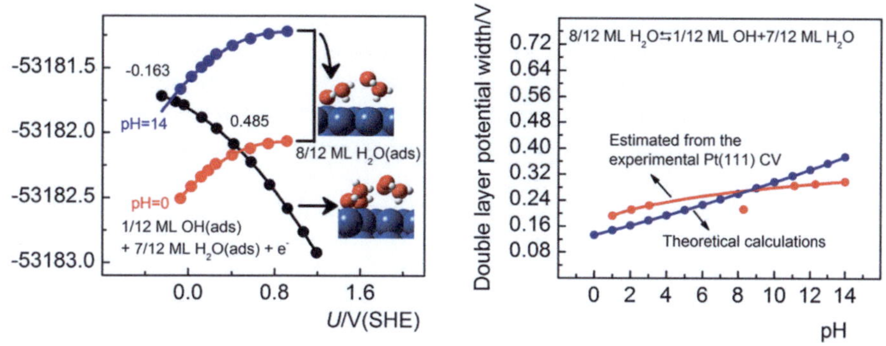


**Fig. 12.7** Left-hand panel gives example of curves used with the along with H(ads) formation potentials to generate pH dependence of double layer widths shown in right-hand panel along with experimental results. As the curve of energy versus potential of the $H_2O$(ads) system is shifted up for pH = 0 to 14, It intersects the OH(ads) system energy to give pH-dependent reversible potentials. For clarity, the $H^+$(aq) and $e^-$ contributions are omitted in the figures. Reprinted with permission from M. Zhao and A. B. Anderson, "Prediction the Double Layer Width on Pt(111) in Acid and Base with Theory and Extracting it from Experimental Voltammograms," J. Phys. Chem. C 121, 28,051–28,064, 2017. Copyright 2017 American Chemical Society

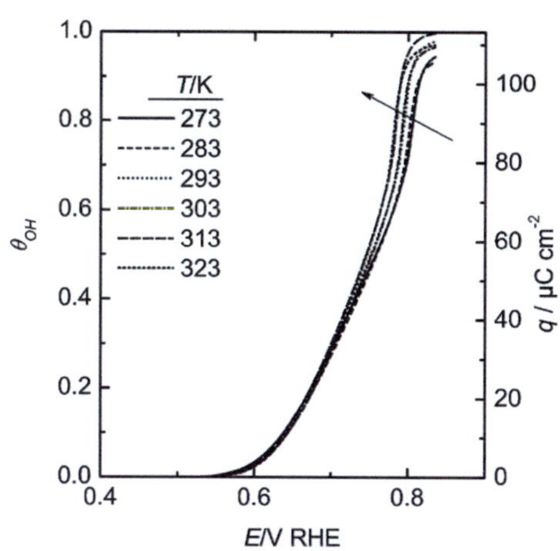


**Fig. 12.8** Charge density of OH adsorption and normalized OH coverage, $\Theta$, on Pt(111) obtained by integration of the voltammograms for pH = 1 (0.1 M $HClO_4$) electrolyte as functions of between 0.4 V and 0.9 V. Six rising curves from 0.5 V to about 0.83 V are for experiments at temperatures from 293 to 323 K in 10 K increments. The arrow indicates the direction of the change associated with the increase of temperature, which is reaching higher OH coverage at lower potentials as the temperature is increased. $\Theta = 1$ corresponds to 0.45 ML of Pt sites covered. Reprinted with permission from V. Climent, R. Gómez J. M. Orts, and J. M. Feliu, "Thermodynamic Analysis of the Temperature Dependence of OH Adsorption on Pt(111) and Pt(100) Electrodes in Acidic Media in the Absence of Specific Anion Adsorption," J. Phys. Chem. B, 110, 11,344–11,351, 2006. Copyright 2006 American Chemical Society

densities at pH 1 that OH(ads) coverage reached a maximum of about 0.45 ML [19]. They found the coverage of OH as a function of potential at temperatures from 273 to 323 K, given in Fig. 12.8. The coverage increases with increasing potential at a fairly constant rate up to site coverage of about 0.25 ML or $\Theta = 0.5$ using the scale in Fig. 12.8. Then the rate increases and the higher rate continues up to the maximum coverage. From the table, the potential for forming 1/2 ML OH(ads) at pH 1 is 0.58 V higher than for forming 1/6 ML. This is qualitatively consistent with the maximum observed coverage of OH(ads) in the first wave being less than 1/2 ML. Then the increase in the slope with increasing potential at around $\Theta = 0.6$ in Fig. 2.8 corresponding to 1/3 ML in Table 1 would result from precipitous deposition of OH(ads) from water oxidation at 0.8 V in pH 1 electrolyte. Voltammograms show the sharp peak decreasing as pH increases [20] Fig. 12.9, taken from this paper illustrates a shape change in the voltammograms. From pH 8.3 to 13.1 there is a single broad peak and between pH 4.3 and pH 8.4 the sharp higher potential peak disappears. The theoretical predictions in Table 12.1 show the two potential regions converging in qualitative agreement with the measurements. At pH 1.2 the onset potential for forming OH(ads) is about 0.5 V and for pH 13.1 it is about 0.2 V. Calculated values from the table are in good agreement, 0.6 V at pH 1 and 0.1 V at pH 14. However, in the low pH region the theory predicts the potential shift needed to form 1/2 ML to 2.3 ML OH(ads) after forming 1/3 ML OH(ads) is ~ 0.6 V which is larger than the ~ 0.3 V width of the whole peak in the voltammogram. This disagreement indicates a need for further refinement of the theoretical model.

When the potential is cycled still higher, a new wave caused by a second oxidation emerges, commencing at about 1.02 V (RHE) at pH 1, shown in the first panel of Fig. 12.10 [21] This wave rises sharply without a low-potential shoulder like for the first oxidation. Calculated data in Table 12.1 include the potentials for further oxidation of $H_2O$(ads) to OH(ads) at pH 1, going from 1/3 ML to 1/2 ML and 1/2 ML to 2/3 ML; both are 1.26 V. Theoretical values for OH(ads) oxidation to O(ads)

**Table 12.1** Calculated reversible potentials (*vs.* SHE) for 1/6 ML,1/3 ML, 1/2 ML, and 2/3 ML $OH_{(ads)}$ formation from pH = 0, 1, 2.05, 3.05, 8.3, 11.1, 12.3, and 14. Reprinted with permission from M. Zhao and A. B. Anderson, "Prediction the Double Layer Width on Pt(111) in Acid and Base with Theory and Extracting it from Experimental Voltammograms," J. Phys. Chem. C 121, 28,051–28,064, 2017. Copyright 2017 American Chemical Society

| Onsets of X ML OH | $U_{rev}$/V(SHE) | | | | | | | |
|---|---|---|---|---|---|---|---|---|
| | pH 0 | pH 1 | pH 2.05 | pH 3.05 | pH 8.3 | pH 11.1 | pH 12.3 | pH 14 |
| 1/6 (~1/3)* | 0.643 | 0.598 | 0.550 | 0.503 | 0.241 | 0.091 | 0.024 | −0.073 |
| 1/3 (~2/3)* | 0.634 | 0.569 | 0.503 | 0.440 | 0.130 | −0.023 | −0.086 | −0.173 |
| 1/2 (~1)* | 1.265 | 1.175 | 1.081 | 0.992 | 0.525 | 0.277 | 0.172 | 0.022 |
| 2/3(~1 1/3)* | 1.256 | 1.170 | 1.079 | 0.993 | 0.536 | 0.290 | 0.184 | 0.033 |

* Numbers in parentheses correspond to coverages quoted in the paper by Climent et al. Their values are greater because in their work maximum coverage was 0.45 of surface Pt atoms which they defined as 1 ML

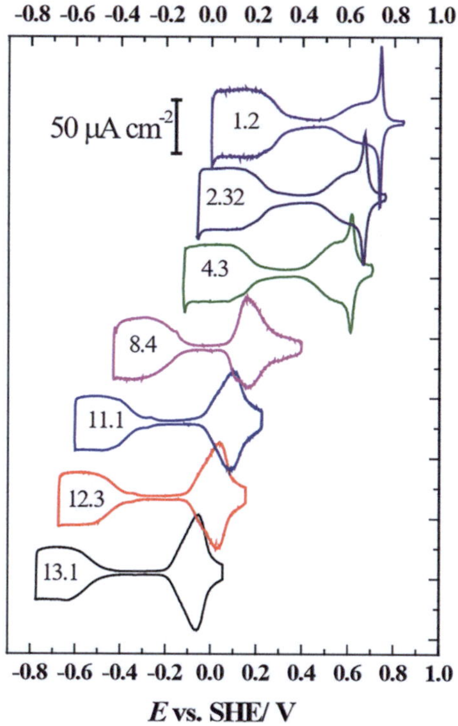


**Fig. 12.9** Cyclic voltammograms over Pt(111) at different values of pH. Reprinted from Electrochim. Acta 162, R. Rizo, E. Sitta, E. Herrero, V. Climent, and J. M. Feliu, Toward the understanding of the interfacial pH scale at Pt(111) electrodes, Electrochim. Acta 162, 138–145, 2015. Copyright 2015, with permission from Elsevier

the second peak have been calculated to be about 1.1 V for pH 1 [22]. Given the uncertainties of the calculated potentials, the closeness of these OH(ads) and O(ads) formation potentials allows speculating that both contribute to the sharp peak in the second wave.

When the pH is 13, the second oxidation wave is broadened and moved slightly lower in potential, closer to first wave. The basic electrolyte appears to interfere with the relatively orderly phase-like transition observed in acid electrolytes.

The presence of the two calculated groups of reduction potentials for $H_2O$(ads) oxidation to OH(ads) in Table 12.1 is easily related to the number of strong hydrogen bond between the adsorbed water molecules and OH(ads) and between the adsorbed hydroxyls. Consider the calculated structures for the hexagonal arrangements of $H_2O$(ads) with OH(ads) in Fig. 12.11. For each 1/6 ML conversion of $H_2O$(ads) to OH(ads) there is a change in surface system energy caused by the difference in $H_2O$ to OH adsorption energies. Upon breaking a strong OH bond in weakly adsorbed $H_2O$(ads), a new strong OH(ads) bond forms to the surface. This in 1/6 ML steps. The reversible potentials in Table 12.1 are only slightly different for forming 1/6

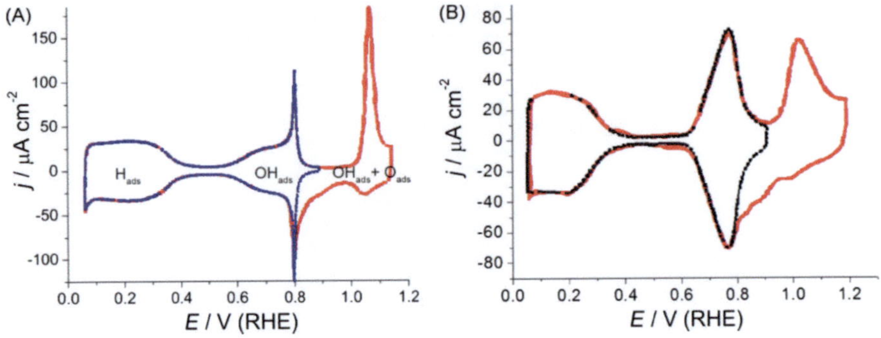


**Fig. 12.10** Left panel (A), stable voltammetric profile of a well-ordered Pt(111) electrode in 0.1 M $HClO_4$ scanned at 50 mV·s$^{-1}$, at two upper potential limits, first 0.9 V (blue) associated with OH(ads) formation showing sharp peaks at about 0.8 V for both the forward and reverse scan, indicating reversibility. The second scan continues to 1.15 V (red) and is associated with forming O(ads) and only the peak of the forward scan is sharp. Right panel (B) is for measurements in 0.1 M NaOH and shows the broadened current density peaks for the OH(ads) in the forward and reverse scans and for O(ads) only the forward scan has a broadened peak. This is Fig. I reprinted from A. M. Gómez-Marín, R. Rizo, and J. M. Feliu, *Beilstein J. Nanotechnol.* **2013,** *4,* 956–967

and 1/3 ML OH(ads) which means there is an approximately constant change in bond strengths as the OH(ads) coverage increases from 1/6 ML to 1/3 ML. The same holds comparing the reversible potentials for forming 1/2 and 2/3 ML OH(ads) though according to Table 12.1 the values are both shifted positive by about 0.6 V when the pH is low. Why are there two potential groups when the pH is low? It is because of the total number of hydrogen bonds. By counting hydrogen bonding interactions in Fig. 12.11, three are seen per OH(ads) for 1/6 and 1/3 ML OH(ads) structures. The count decreases to two for 1/2 and 2/3 ML OH(ads) and the calculated jump to higher reversible potential between 1/3 and 1/2 ML OH(ads) is attributed to this loss in stability. The jump in oxidation potential is understood as due to the loss of hydrogen bonding stabilization in the oxidation product. Other theoretical work employing lower net surface coverage models found similar relationships between the reversible potentials for $H_2O$(ads) oxidation and the number of hydrogen bonds [23].

The current densities for the second oxidation wave in Fig. (12.10) and the corresponding reduction wave do not have mirror symmetry like the first wave. This means the second oxidation is irreversible. When the potential is cycled back, the shape of the reduction wave means that to reduce of some surface species requires going to lower potentials. This implies that at potentials > 1.0 V and that the putative Pt(O,OH) oxidation products relax into lower energy states following electron transfer, possibly accompanied by metal atom displacements from their original positions. According to experimental literature, they may fall into Gibbs energy stability wells with increased entropy and probably liberating heat [24]. To undo these bonds requires higher energy electrons and these are provided when the applied potential is lowered, raising $E_f$. If the reduced Pt surface is left restructured at first, with

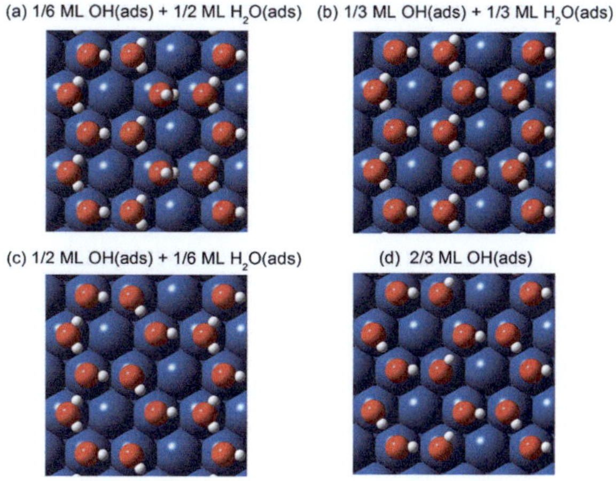


**Fig. 12.11** Normal view of $H_2O$(ads) and OH(ads) at coverages indicated. Reprinted with permission from M. Zhao and A. B. Anderson, "Prediction the Double Layer Width on Pt(111) in Acid and Base with Theory and Extracting it from Experimental Voltammograms," J. Phys. Chem. C 121, 28,051–28,064, 2017. Copyright 2017 American Chemical Society

annealing it will revert to the original ordered surface. Further discussion is in the paper of Gómez-Marín et al. where essentially the same conclusions are presented for the oxidation waves in acid solution.

## 12.9 Further Experimental and Theoretical Approaches to the Understanding the Water Oxidation Voltammogram

Earlier theoretical studies of the oxidation of $H_2O$(ads) to OH(ads) and O(ads) have invoked dynamic Monte-Carlo(MC) calculations within the LRC model for activation energies [25]. Another analysis employed the CHE model calculating interaction and formation energies including an electric field term [26]. These modeling studies fairly accurately predicted onset potentials and broad voltammogramic features for OH(ads) and O(ads) coverage up to about 1.0 V where predictions were terminated due to limitations of the description of adsorbate interactions. Visvanathan et al. showed predicted coverages of $H_2O$(ads), OH(ads) and O(ads) were in qualitative agreement with interpretations of ex situ XPS measurements for emersed Pt(111) electrodes [27]. Graphs of coverage of $H_2O$, OH and O versus potential are in Fig. 12.12. The electrolyte for the measurements was 0.1 M HF, with pH 1 and the theoretical studies employed intermediates generated by acid mechanisms. Both the experimental and theoretical curves lack the detail present in the in situ cyclic voltammograms discussed above.

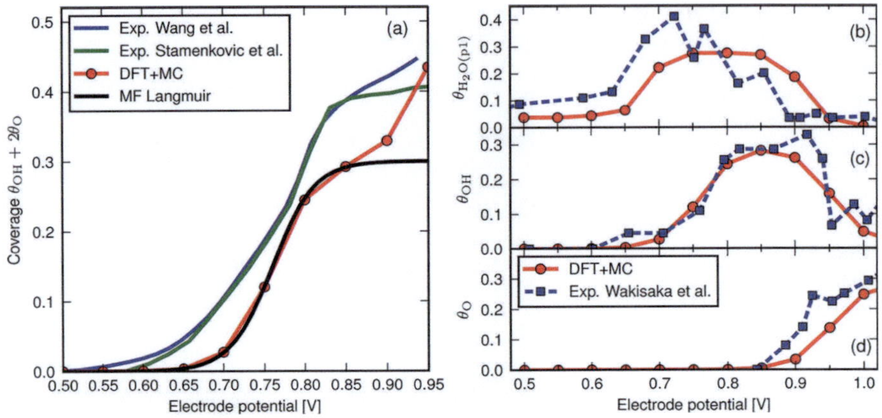


**Fig. 12.12** From Visvanathan et al. The left-hand panel shows total coverage by oxygen species from experiments of Wang. Et al; Stamenkovic et al. DFT + MC by Rai et al. and Langmuir. The first begin increasing at about 0.5 V and 0.6 V and the latter two at 0.65 V. The Langmuir curve reaches a maximum and levels off at 0.3 coverage at 0.85 V while the others continue to about 0.4 coverage without leveling off. The right-hand panel has three sub-panels and they compare coverage predictions from DFT + MC calculations with coverages estimated from XPS excite measurements of Wakisaka et al. Results for $H_2O$(ads) coverage (b), OH(ads) coverage (c) and O(ads) coverage (d) show rough agreements between theory and experiment. The largest disagreement is the underestimate by theory for $H_2O$(ads) below 0.75 V and its overestimate above 0.9 V. This is Fig. 12.3 from V. Visvanathan, H. A. Hansen, J. Rossmeisl, T. F. Jaramillo, H. Pitsch, and J. K. Nørskov, "Simulating Linear Sweep Voltammetry from First-Principles: Application to Electrochemical Oxidation of Water on Pt(111) and $Pt_3Ni$(111)," J. Phys. Chem. C 116, 4698–4704, 2012

## 12.10 Deviations from Pourbaix Diagrams in the Case of Reacting Surfaces

The pH dependencies of reversible potentials for reactions in aqueous solutions follow the Nernst equation. Pourbaix diagrams on the other hand are phase diagrams depicting the equilibrium potentials between, for example, metal and oxide phases as functions of pH. Starting with the general reaction for aqueous reactants and products,

$$\mathrm{Ox} + \mathrm{mH}^+ + \mathrm{ne}^- \rightleftharpoons \mathrm{Red} \tag{12.10}$$

where the product may include water molecules, the Nernst equation is

$$U_{\mathrm{rev}} = U^0 + (0.0592/\mathrm{n})[\mathrm{Ox}]/[\mathrm{Red}] - (\mathrm{m/n})(0.0592\mathrm{pH}) \tag{12.11}$$

Equation (12.11) says the reversible potential for Eq. (12.10) varies linearly with pH when Ox and Red concentrations are fixed. This is unambiguous when the equilibrium is between an aqueous phase of constant concentration and a solid phase or two solid phases with unit activity. How about at an interface?

Let us label the linearity of $U_{rev}$ versus pH as the solid/solution behavior. Theoretically calculated reversible potentials versus pH plots in Fig. 12.4 show that there can be notable deviation from linearity in graphs of $U_{rev}$ for reactions on and including surfaces. When going from one coverage of OH(ads) to a higher coverage, such as 0 to 1/6 ML shown above, this is caused by the potential dependencies of the adsorption energies. In this case 2/3 ML $H_2O$(ads) on Pt(111) is oxidized to 1/2 ML $H_2O$(ads) + 1/6 ML OH(ads). More examples from theoretical calculations are available [28]. Reference [28] contains references to literature where the adsorption energies are assumed independent of potential so that the Pourbaix diagrams for surface reactions follow Eq. (12.12). Those works employ the CHE/LGER model where potential is inserted only for the

$$\mathrm{Ox}(U) + \mathrm{mH}^{+}(\mathrm{aq}) + \mathrm{ne}^{-}(U) \rightleftharpoons \mathrm{Red}(U) \tag{12.12}$$

electron energy in Eq. (12.8) but not for adsorption energies. Including dependencies of adsorbate energies for all species gives an equation similar to Eq. (12.8) in place of Eq. (12.11). The reaction in Eq. (12.12), when processed through Eq. (12.8), provides the $U_{rev}$ versus pH plots for partial oxidations of the surface. Again, as Fig. 12.4 shows, they are not necessarily linear or with the slope dictated by solid/solution Pourbaix diagrams.

## 12.11 What Deviations from Pourbaix Diagrams for Surface Reactions Mean to Adsorption Bond Strengths

The deviations mean that as functions of solution pH, the onset potentials for metal oxidation reactions may be higher or lower than calculated by application of the Nernst equation for the bulk oxide. If the initial oxide or hydroxide layer product is particularly stable relative to the bulk oxide, then the potential of the onset of oxidation current will be less than expected from the Nernst equation for the solid–liquid interface. An example is the above platinum oxidation: as the potential increased, the coverage of OH(ads) and then O(ads) increased and a layer of platinum oxide began to form. Since coverage increases at higher potentials, to maintain equilibrium at each incremental increase in coverage the adsorption energy of OH or O becomes less. That is, as the electron becomes more stable (more negative in energy) at higher potentials the adsorbate–surface bond becomes less stable (less negative in energy). This is seen in voltammograms like Fig. (8): as OH(ads) coverage increases the reversible potential for incremental increases in its coverage also increases because the adsorption energy or reaction energy is decreasing. The water molecules that are present, some at the interface becoming oxidized, also contribute to the energy of reaction. Because the hydrogen bonds are relatively weak, the OH and O bond strengths to the surface dominate the changes in reversible potential as their coverage increases. When the surface takes bulk oxide characteristics, the usual Pourbaix equation describes the equilibrium.

## References

1. Zhang T, Anderson AB (2007) Oxygen reduction on platinum electrodes in base: theoretical study. Electrochim Acta 53:982–989
2. Conway BE, Wilkinson DP (1993) Non-isothermal cell potentials and evaluation of entropies of ions and of activation for single electrode processes in non-aqueous media. Electrochim Acta 38:997–1–13
3. For example, Anderson AB, Roques J, Mukerjee S, Murthi VS, Markovic NM, Stamenkovic V (2005) Activation energies for oxygen reduction on platinum alloys: theory and experiment. J Phys Chem B 109:1198–1203
4. Kastlunger G, Wang L, Govindarajan N, Heenen HH, Ringe S, Jaramillo T, Hahn C, Chan K (2022) Using pH dependence to understand mechanisms in electrochemical CO reduction. ACSCatal. 12:4344–4357
5. Zhao M, Anderson AB (2016) Predicting pH dependencies of electrode surface reactions in electrocatalysis. Electrochem Comm 69:64–67
6. Cheng J, Liu X, VandeVondele J, Sulpizi M, Sprik M (2014) Redox potentials and acidity constants from density functional theory based molecular dynamics. Acc Chem Res 47(12):3522–3529
7. Uddin J, Anderson AB (2013) Trends with coverage and pH in stark tuning rates on Pt(111). Electrochim Acta 108:398–403
8. Stamenkovic V, Chou KC, Somorjai GA, Ross PN, Markovic NM (2005) Vibrational properties of CO at the Pt(111)-solution interface: the formation of anomalous Stark Tuning slope. J Phys Chem B 109:678–680
9. Chang SC, Weaver MJ (1990) Coverage-dependent dipole coupling for carbon monoxide adsorbed at ordered Pt(111)-aqueous interfaces: structural and electrochemical implications. J Chem Phys 92:4582–4594
10. Zhao M, Anderson AB (2017) Prediction the double layer width on Pt(111) in acid and base with theory and extracting it from experimental Voltammograms. J Phys Chem C 121:28051–28064
11. Asiri HA, Anderson AB (2013) Using gibbs energies to calculate the Pt(111) $H_{upd}$ Cyclic Voltammogram. J Phys Chem C 117:17509–17513
12. Ojha K, Doblhoff-Diera K, Koper MTM (2022) Double-layer structure of the Pt(111)–aqueous electrolyte interface. PNAS 119:1–9
13. Shin S-J, Kim DH, Bae G, Ringe S, Choi H, Lim H-K, Choi CH, Kim H (2022) On the importance of the electric double layer structure in aqueous electrocatalysis. Nat Commun 13:174
14. Haq S, Harnett J, Hodgson A (2002) Growth of thin crystalline ice films on Pt(111). Surf Sci 505:171–182
15. Glebov A, Graham AP, Menzel A, Toennies JP (1997) Orientational ordering of two-dimensional ice on Pt(111). J Chem Phys 106(22):9382–9385
16. Morgenstern M, Michely T, Comsa G (1996) Anisotropy in the adsorption of $H_2O$ at low coordination sites on Pt(111). Phys Rev Lett 77:703–706

17. Morgenstern M, Muller J, Michely T, Comsa G (1997) The Ice Bilayer on Pt(111): nucleation, structure and melting. Z Phys Chem 198:43–72
18. Doering DL, Madey TE (1982) The adsorption of water on clean and oxygen-dosed Ru(001). Surf Sci 123:305–337
19. Climent V, Gómez R, Orts JM, Feliu JM (2006) Thermodynamic analysis of the temperature dependence of OH adsorption on Pt(111) and Pt(100) electrodes in acidic media in the absence of specific anion adsorption. J Phys Chem B 110:11344–11351
20. Rizo R, Sitta E, Herrero E, Climent V, Feliu JM (2015) Toward the understanding of the interfacial pH scale at Pt(111) electrodes. Electrochim Acta 162:138–145
21. Gómez-Marín AM, Rizo R, Feliu JM (2013) Some reflections on the understanding of the oxygen reduction reaction at Pt(111). Beilstein J Nanotechnol 4:956–967
22. Jinnouchi R, Anderson AB (2008) Aqueous and surface redox potentials from self-consistently determined gibbs energies. J Phys Chem C 112:8747
23. Tian F, Jinnouchi R, Anderson AB (2009) How potentials of zero charge and potentials for water oxidation to OH(ads) on Pt(111 electrodes vary with coverage. J Phys Chem C 113:17848–17892
24. Markovic´ NM, Schmidt TJ, Grgur BN, Gasteiger HA, Behm RJ, Ross PN (1999) Effect of temperature on surface processes at the Pt(111)-liquid interface: hydrogen adsorption, oxide formation, and CO oxidation. J Phys Chem B 103:8568–8577
25. Rai V, Aryanpour M, Pitsch H (2008) First-principles analysis of oxygen-containing adsorbates formed from the electrochemical discharge of water on Pt(111). J Phys Chem C 112:9760–9768
26. Visvanathan V, Hansen HA, Rossmeisl J, Jaramillo TF, Pitsch H, Nørskov JK (2012) Simulating linear sweep voltammetry from first-principles: application to electrochemical oxidation of water on Pt(111) and Pt3Ni(111). J Phys Chem C 116:4698–4704
27. Wakisaka M, Suzuki H, Mitsui S, Uchida H, Watanabe M (2009) Identification and quantification of oxygen species adsorbed on Pt(111) single-crystal and polycrystalline Pt electrodes by photoelectron spectroscopy. Langmuir 2009(25):1897–1900
28. Anderson AB (2018) Pourbaix diagrams for $H_2O$ oxidation to adsorbed OH on Pt(111) and why they differ from those for bulk solids. J Phys Chem C 122:9958–9964

# Chapter 13
# Concluding Comments and the View Ahead

**Abstract** A whirlwind historical summary of the phenomenological and quantum based concepts and computational methods developed for electrochemistry in this book is followed the suggestion that machine learning and artificial intelligence may open a new frontier in prediction and application.

This book led you through the evolution in conceptual models used to interpret measurements and make predictions and provide understanding in the science of electrochemistry. Models are constructs in our minds and are similar to, if not quite the same as, theories. When considering the material in scientific talks, papers, and books one should do so from the perspective expressed by Faraday in Chap. 1: "All theories are fixed upon uncertainties and all of them want alteration and support."

Our story began in the 1800s with the observations of several of the famous historical names of electrochemistry, Volta, Davies, Nicholson, and Faraday. Cell voltages and current rates and amounts were measured. Batteries were constructed and used for metal plating and gas evolution reactions. Faraday discovered electrical and chemical equivalents and Thompson showed electric currents are comprised of the motions of charged particles now called electrons.

Going into the twentieth century, Nernst established the relationship of cell potential to reactant ion concentration in the electrolyte. Gibbs constructed the science of thermodynamics and showed the maximum amount of electrical work a battery could provide was proportional to the potential drop the electrons experience and the current passed. More generally, he discovered classical thermodynamics, which lay in wait for upcoming quantum applications using Schrödinger's wave equation for matter.

The focus during the early 1900s was on the kinetics of electrochemical reactions. From the dependencies of measured currents on electrode potentials, Tafel, Butler, and Volmer established mathematical formulas connecting current and potential. Since then, the fitting of measurements using these mathematical formulas became for decades the approach for modeling and understanding electrochemistry. Although there were simplifying physical assumptions behind the mathematical developments, they did not prevent the development of a useful framework for understanding electrochemistry that still applies today. Engineering and development applications for



A. Anderson, *Past and Current Models in Electrochemistry*,
https://doi.org/10.1007/978-3-031-97791-6_13

batteries, power, plating, and synthesis thrived and continue to thrive based on this level of understanding. In other words, first or second order extrapolations from zeroth order state descriptions are often good enough. Much of the theoretical development in this time period consisted in mathematical analysis involving extrapolation into relatively unknown or unexplored domains.

The discovery of the Schrödinger wave equation uncovered a new horizon for exploration. At first, predictions from computations using quantum mechanics played a minor role, being unreliable or not applicable to realistic models of chemical systems because computers at the time lacked the storage capacity and processor speed needed for more advanced calculations. Surface sensitive spectroscopies did not yet exist, meaning surface compositions and structures could only be guessed. These limitations postponed until recent years the development of more detailed understanding of actual structures, compositions, and changes at the electrochemical interface. The classical theories of Marcus and Hush, with their focus on classical solvation energies and Born–Oppenheimer vibrational potentials, gave insights to electron transfer processes during the interim.

Using the Schrödinger equation and the free electron Hamiltonian model, Gurney set groundwork for developing the modern models for the electrochemical interface. He introduced densities of states for metallic electrodes and assigned their electron chemical potentials, $\mu_e$, equal to the negative of workfunctions of the metals in solution plus a constant. When changing the applied potential, $\mu_e$ of the surface electrons and the surface workfunction changed proportionately. It was understood this happened when charge was added to the surface, triggering a charge-compensating layer of electrolyte ions to form. Gurney employed classical vibrational potentials to model outer-sphere electron transitions and Butler included a potential of interaction with the surface to model inner-sphere reactions.

Late in the twentieth century a simplified model was used to calculate the response of properties of molecules adsorbed on an electrode to changing the applied voltage. In its application the electrode valence band was shifted, by using a parameterized model Hamiltonian, parametrically over a range of energies. The Hamiltonian for this semiempirical band shift method contained no explicit electron repulsion components and the electronic structure was changed "by hand" to accommodate changes in the Fermi energy, $E_f$, and $\mu_e$ with changes in applied electrode potential. In these studies, the original Gurney picture of the electrode surface valence band moving up and down on the energy scale as the applied potential changed was implemented without including a model for the double layer. The extended Hückel (EH) Hamiltonian could model potential changes this way and, with two-body repulsion energies added to the sum of occupied orbital energies as in the atom superposition and electron delocalization molecular orbital (ASED-MO) procedure, estimates of bond strengths and electrochemical vibrational tuning rates (EVTR) were calculated. The ability to predict the tuning rate trends by calculating the frequencies as functions of surface band position without the presence of an electric field meant the tuning rate was not necessarily influenced by the classical Stark effect. Research today is evaluating the relative contributions of potential-dependent inductive effects and double layer electric field effects. Energies calculated for electron transfer reactions using ASED-MO were not sufficiently accurate for anything more than qualitative discussions.

Early in the twenty-first century new models were developed that ignored the dependence of the electrode $E_f$ on potential and had the reacting electron transfer from a fictitious source with a specified $\mu_e$ but existing outside the reacting system. This made the electron affinity or ionization potential of the reaction center and their dependence on structure the focus. The electrode potential was assigned a value relative to the hydrogen electrode which is 0 V with a thermodynamic workfunction of about 4.43 eV. Reaction energies, activation energies and reversible potentials predicted with this model omitted the perturbations on adsorbate–surface bonds resulting from the $E_f$ shifts that accompany voltage changes. In other words, all calculations were performed at the potential of zero charge. A limitation was that model for the electrode surface was restricted to one or two metal atoms. However, the local nature of chemical bonds meant the results of calculations with the model were meaningful.

The types of pzc-based models presented in this book can be split into classical and quantum mechanical. The classical branch included the Butler-Volmer theory, the Hush theory, and the Marcus theory and its extensions. The quantum mechanical branch includes the semiempirical EH and ASED-MO procedures and also the local reaction center (LRC) model, the computation hydrogen electrode (CHE), and the linear Gibbs energy relationship (LGER) model. The latter three employ full Hamiltonians and self-consistent-field (SCF) calculations but have no provision for the changes in $E_f$ that occur in response to potential changes. Any of them could be used as zeroth order starting points for attempting making improve models. For example, it is possible to include solvating molecules in the band shift method but this was not pursued. Efforts have been made to include solvation in SCF calculations. However, inaccuracy in calculated activation energies and reversible potentials is always present when potential-dependent polarizations of the chemical bonds between reactants and products and the electrode surface are not included in the calculations.

At the same time, other approaches were explored, beginning with bathing cluster models of the electrochemical interface in electric fields generated by assigned charge distributions and calculating vibrational Stark tuning rates with ab initio quantum mechanics. This was followed by DFT band calculations that employed translational cells with electrons added to or subtracted to them and counter charges distributed according to models of various types with the goal of calculating electron transfer activation energies. For a given charge added to the translational cell to define an initial energy, $E_f$ was not fixed during the course of a reaction in these studies and so errors in calculated activation energies and reaction potentials, though hopefully small, were indeterminate.

Recently written DFT codes can add fractional charges to translation cells and include the necessary compensating charge described with statistical theory in the double layer. Upon adding a layer or two of solvating water molecules, these codes predict relatively accurate reaction reversible potentials. Calculating transition states and activation energies at a given potential is computationally intensive because of the large number of structure and energy points needed. These codes are now the starting point for quantitative analysis and predictions of structures, potentials, and activation energies at the electrochemical interface. Typically, several layers of atoms are used to represent the electrode surface in these calculations, which generally provide accurate descriptions of the bonding and reaction chemistry on the surface. The added charge shifts $\mu_e$ for the slab and the fact that the bulk of the electrode material beneath it has a different $\mu_e$ is accommodated with the concept of band bending which is a way after accounting for the charge polarization at the interface between the surface slab and the bulk metal. The bending corresponds to a smooth transition of the electronic structure from the surface to the bulk. The electron in an oxidation reaction enters the bottom of the conduction band at the surface and remains there, passing through interface polarizations, until emerging at the cathode surface to participate in a reduction reaction. The work performed by the electron and voltage between the anode and cathode half-cells is not affected by polarizations at interfaces along the way.

The two broad goals of theoretical modeling will continue to be (i) understanding and prediction for the general audience and (ii) engineering and development for commercial applications. For the academic viewpoint most efforts involve current–voltage measurements, spectroscopic measurements, and their theoretical understanding within the models and quantum mechanical calculations of structures and energy states have been the players so far. From the data obtained, modified or new systems are proposed and tested. Trends are found, analyzed, and understood. However, the model systems used in the experimental and computational studies are only approximations or guesses when compositions and structures at electrochemical interfaces are not fully known, which is often the case. The engineering and commercial developers will continue testing new catalysts and synthesis reactions and will continue optimizing activities and stabilities of existing processes.

Artificial intelligence (AI) and machine learning (ML) are finding their way into electrochemistry. These techniques, applied using the growing experimental and theoretical data bases, will be useful by providing good proposals for new electrocatalysts and new and more efficient industrial processes.

The application of AI techniques to electrochemical, materials, and to the physical and biological sciences generally, is likely to be revolutionary. Young science students need to learn all they can about AI in their field of interest. Universities need to do all they can to assist.

# Index of Names




A. Anderson, *Past and Current Models in Electrochemistry*,
https://doi.org/10.1007/978-3-031-97791-6

# Index of Topics





A. Anderson, *Past and Current Models in Electrochemistry*,
https://doi.org/10.1007/978-3-031-97791-6

## E



## F



## G



## H

MIX
Papier aus verantwortungsvollen Quellen
Paper from responsible sources
FSC® C105338

If you have any concerns about our products,
you can contact us on
**ProductSafety@springernature.com**

In case Publisher is established outside the EU,
the EU authorized representative is:
**Springer Nature Customer Service Center GmbH**
**Europaplatz 3, 69115 Heidelberg, Germany**

Printed by Libri Plureos GmbH
in Hamburg, Germany